Study Guide

INTRODUCTION TO PROBABILITY
AND STATISTICS, 4th Edition

Study Guide

INTRODUCTION TO PROBABILITY AND STATISTICS, 4th Edition

William Mendenhall

Prepared by

Barbara M. Beaver

Robert J. Beaver
University of California, Riverside

DUXBURY PRESS
A division of Wadsworth Publishing Company, Inc.
North Scituate, Mass. · *Belmont, California*

DUXBURY PRESS

A DIVISION OF WADSWORTH PUBLISHING COMPANY, INC.

L.C. Cat. Card No.: 74–84837
ISBN 0–87872–095–2

PRINTED IN THE UNITED STATES OF AMERICA

3 4 5 6 7 8 9 10 — 79 78 77

CONTENTS

PREFACE

The study of statistics differs from the study of many other college subjects. One must not only absorb a set of basic concepts and applications but must precede this with the acquisition of a new language.

We think with words. Hence, understanding the meanings of words employed in the study of a subject is an essential prerequisite to the mastery of concepts. In many fields, this poses no difficulty. Often, terms encountered in the physical, social, and biological sciences have been met in the curricula of the public schools, in the news media, in periodicals, and in everyday conversation. In contrast, few students encounter the language of probability and statistical inference before embarking on an introductory college-level study of the subject. Many consider the memorization of definitions, theorems, and the sytematic sequence of steps necessary for the solution of problems to be unnecessary. Others are oblivious to the need. The consequences for both types of students are disorganization and disappointing achievement in the course.

This study guide attempts to lead the student through the language and concepts necessary for a mastery of the material in *Introduction to Probability and Statistics,* 4th edition, by William Mendenhall (Duxbury Press, 1974).

A study guide with answers is intended to be an individual student study aid. The subject matter is presented in an organized manner that incorporates continuity with repetition. Most chapters bear the same titles and order as the textbook chapters. Within each chapter, the material both summarizes and reexplains the essential material from the corresponding textbook chapter. This allows the student to gain more than one perspective on each topic and, hopefully, enhances his understanding of the material.

At appropriate points in each chapter, the student will encounter a set of Self-Correcting Exercises in which problems relating to new material are presented. Terse, stepwise solutions to these problems are found at the back of the study guide. These can be referred to by the student at any intermediate point in the solution of each problem, or used as a stepwise check on any final answer. The Self-Correcting Exercises not only provide the student with the answers to specific problems, but also reinforce the stepwise logic required to arrive at a correct solution to each problem.

At the end of each chapter, additional sets of exercises can be found. These exercises are provided for the student who feels that further individual practice is needed in solving the kinds of problems found within each chapter. At this point, having been given stepwise solutions to the Self-Correcting Exercises, the student is now presented with only final answers to problems. Hopefully when the student's answer disagrees with that given in the study guide, he should be able to find his error by recalculating and comparing his solution with the solutions to similar Self-Correcting Exercises. If the answer given disagrees with the student's only in decimal accuracy, it can be assumed that this difference is due only to rounding error at various stages in the calculations.

When the study guide is used as a supplement, the textbook chapter should be read first. Then the student should study the corresponding chapter within the study guide. Key words, phrases, and numerical computations have been left blank for the student to insert his response. The answers are presented in the page margins. These should be covered until the student has supplied his response for each blank. One should bear in mind, though, that in some instances more than one answer is appropriate for a given blank. It is left to the reader to determine whether his answer is synonomous with the answer given within the margin. Since perfection is something to be desired, we ask that the reader who has located an error kindly bring it to our attention.

Barbara M. Beaver
Robert J. Beaver

Chapter 1

INTRODUCTION

1.1 The Objective of Statistics

Statistics involves sampling from a larger body of data called a population. For example, the gubernatorial preferences of eligible voters in an election represent a population of interest to the politicians of a state. An inference about the characteristics of this population can be obtained from information contained in a sample. The inference will involve either an estimate or decision concerning the characteristics of the population.

A medical experiment was conducted to determine the absorption of a drug in the heart of a rat. A fixed amount of the drug was injected into each of ten rats, the rats were killed, and the amount of drug absorbed by each heart measured. The measurements on hearts of the ten rats represent a sample drawn from a conceptual _____ of measurements on the very large number of rat hearts that might have been obtained under similar experimental conditions. The information contained in the _____ of ten heart measurements is to be used to make inferences about the conceptual population from which the sample was drawn.

population

sample

Statistics is concerned with a theory of information and its application in making inferences based on sample information about populations in the sciences and industry. The objective of statistics is to _____ _____ about a population from information contained in a sample.

make
inferences

1.2 The Elements of a Statistical Problem

We have noted that the objective of statistics is to make inferences about a _____ based on information contained in a _____.
Sampling implies the acquisition of data, so statistics is concerned with a theory of information. The attainment of the objective of statistics—inference making—is dependent upon three steps, which we will call the elements of a statistical problem.

population; sample

The sample contains a quantity of information on which the inference about the population will be based. In fact, information can be quantified as easily as weight, heat, profit, or other quantities of interest. Consequently, the first

sampling
design

step in a statistical problem is deciding upon the most economical procedure for buying a specified quantity of information. This is called the _____ procedure or the _____ of the experiment. The cost of the specified amount of information will vary greatly depending upon the method used for collecting the data in a sample.

sample

The second step in a statistical problem involves the extraction of information contained in the _____. By analogy, suppose that information was measured in units of pounds (which it is not). It is not unusual for an experimenter to extract only three pounds from a sample that contains ten pounds of information. Thus extracting information from a sample is equivalent to the problem of extracting juice from an orange. We wish to obtain the

maximum

_____ amount of information from a given set of data.

inference

The third step in a statistical problem involves the use of the information in a sample to make an _____ about the population from which the sample was drawn. Some inferences, say estimates of the characteristics of the population, are very accurate and consequently, good. Others are far from reality and bad. It is therefore necessary to clearly define a measure of goodness for an inference maker. Most people observe the world about them and make inferences daily. Some of these subjective inference makers are very good and accurate; others are very poor. Statistical inference makers are objective rather than subjective, but they vary in their goodness. The statistician wishes to obtain the best inference maker for a given situation.

A measure of the goodness or reliability of an inference is always necessary to be able to assess its practical value. Thus, inference making is regarded as a two-step procedure. First, we select the best method and use it to make an

inference
reliability

_____. Second, we always give a measure of the goodness or _____ of the inference.

design
analysis
goodness

The three elements of a statistical problem are (1) _____ of the experiment, (2) _____ of the data, and (3) the making of inferences together with a measure of their _____.

Chapter   2

USEFUL MATHEMATICAL NOTATION

2.1 Introduction

To fulfill the objective of statistics, information must be extracted from measurements contained in a sample. Formulas for extracting this information are typically expressed in summation notation. A second useful notation, functional notation, is important because it is used in the development of summation notation and in presenting formulas that we will encounter later in the text. Consequently, Chapter 2 is devoted to familiarizing the student with functional notation and summation notation.

2.2 Function and Functional Notation

A function consists of two sets of elements and a rule which associates one and only one element of the second set with each of the elements in the first set. It can be displayed graphically, showing the elements of the first and second sets as points inside respective enclosures and the rule of association indicated by joining associated points (elements) with lines.

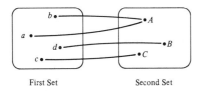

First Set   Second Set

The function visualized above might also be exhibited as a collection of ordered pairs. In each pair, the right member is an element of the second set, and *no two pairs* have the same (first, second) element. Thus our function can also be displayed as the collection $\{(a,A),(b,A),(d,B),(\underline{\hspace{2cm}})\}$ of ordered

first
c,C

first

pairs. Though two of these pairs have the same second element, we note that no two of these pairs have the same _____ element.

A third manner of representing a function is to employ functional notation. The element of the second set corresponding to the element x of the first set is denoted by an expression of the type $f(x)$.

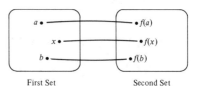

First Set   Second Set

The functional value $f(x)$ is most commonly expressed by a formula when the first and second sets are sets of numbers.

Example 2.1
Write down the collection of ordered pairs represented in functional notation by:

$$f(x) = x^2, x = 1, 2, 3.$$

Solution

3,9

1, 4, 9

$\{(1,1), (2,4) (\underline{\hspace{1.5cm}})\}$. Here the first set is the set of numbers $\{1, 2, 3\}$, while the second set is the set of numbers $\{\underline{\hspace{1.5cm}}\}$.

Example 2.2

$$f(y) = 2y^2 - 3y + 1, \quad y = 0, 1, 2, 3.$$

Supply each of the following functional values:

1

$f(0) = \underline{\hspace{1.5cm}}.$

0

$f(1) = \underline{\hspace{1.5cm}}.$

3

$f(2) = \underline{\hspace{1.5cm}}.$

10

$f(3) = \underline{\hspace{1.5cm}}.$

This function is the collection of ordered pairs:

3,10

$$\{(0,1), (1,0), (2,3), (\underline{\hspace{1.5cm}})\}.$$

Some Useful Functions
Functions of the type $f(x) = (1/2)^x$, $x = 0, 1, 2, \ldots, n$ are used in later chapters. We shall be consistent with mathematical convention and define $a^0 = 1$ if a is not zero. Thus, if $f(x) = (1/2)^x$, then $f(0) = 1$. Also $f(2) = 1/4$

and $f(4) =$ _____ . The letter used to represent a typical value from the
first set (here we have used the letter x) is called the independent variable.
$f(x)$ is then said to be the _____ variable since the value taken on
by $f(x)$ depends on the value used for x.

The factorial function, $f(n) = n!$, $n = 0, 1, 2, \ldots$, is used in the expression
of certain important probability distributions. Again we follow mathematical
convention and define $f(0) = 0!$ to be 1 ($0! = 1$). To complete the definition
of $n!$ we require that $n! = n(n-1)!$ for $n = 1, 2, 3, \ldots$. Hence,

$$1! = (1)(0)! = 1.$$

$$2! = (2)(1)! = 2.$$

$$3! = (3)(2)! = 3 \cdot 2 \cdot 1 = \underline{\hspace{1cm}}.$$

$$4! = (4)(3)! = 4 \cdot 3 \cdot 2 \cdot 1 = \underline{\hspace{1cm}}.$$

and so on.

Example 2.3
The student should not be disturbed if letters other than x are used to represent independent variables or if symbols other than $f(x)$ are used to represent
dependent variables. Supply each missing entry in the following table:

Value of Independent Variable	Formula for Dependent Variable	Value of Dependent Variable
a. $x = 2$	$f(x) = x^2 - 1$	$f(2) =$ ____
b. $y = -1$	$g(y) = 2y + 1$	$g(-1) =$ ____
c. $x = 2$	$f(x) = 3^x - 1$	$f(2) =$ ____
d. $t = 0$	$f(t) = 3^t - 1$	$f(0) =$ ____
e. $u = 3$	$h(u) = u!$	$h(3) =$ ____
f. $u = 0$	$h(u) = u! + 1$	$h(0) =$ ____
g. $i = 5$	$g(i) = x_i$	$g(5) =$ ____

Self-Correcting Exercises 2A

1. $f(x) = 2x + 3$. Find $f(4)$.
2. $g(t) = t^2 - 2$. Find $g(0)$ and $g(4)$.
3. $h(x) = x!$ Find $h(0)$ and $h(4)$.
4. $g(u) = u!$ Find $g(0) + g(4)$.

2.3 Numerical Sequences

If the functional values corresponding to the values $y = 1, y = 2, \ldots$ for the
independent variable are arranged so that $f(1)$ is in the first position, $f(2)$ is in
the second position, $f(3)$ is in the third position, and so on, we say that the
ordered array

margin answers: 1/16, dependent, 6, 24, 3, -1, 8, 0, 6, 2, x_5

18

3(3)

3y

3

3y + 2

3y + 2

$$\{f(1), f(2), f(3), \ldots\}$$

is a sequence. To select or identify a specific term in a sequence, we need only specify its position in the sequence. For example, if the first four terms of a sequence are 3, 6, 9, 12, the fifth and sixth terms are 15 and _____, respectively. When the elements of a sequence correspond to the functional values of $f(y)$ for $y = 1, 2, 3, \ldots$, then y is often called a position variable, since the value given to y determines the position of the element $f(y)$ in the sequence.

Example 2.4

Find a formula expressing the typical element of the sequence

$$3, 6, 9, 12, \ldots$$

as a function of its position in the sequence.

Solution

We note that the element in the first position is 3(1), the element in the second position is 3(2), the third is _____, and so on. Hence, a formula expressing the typical element as a function of its position is

$$f(y) = \text{_____} \quad y = 1, 2, 3, \ldots$$

Example 2.5

Find a formula expressing the typical element of the sequence

$$5, 8, 11, 14, \ldots$$

as a function of its position in the sequence.

Solution

1. One might notice that the terms of this sequence appear to be related to the sequence in Example 2.4, since $5 = 3(1) + 2$, $8 = 3(2) + 2$, $11 = 3(\underline{\hspace{1cm}}) + 2$, and so on. Hence we could use the function

$$f(y) = \text{_____} \quad y = 1, 2, 3, \ldots$$

2. But the sequence could also be described by considering $5 = 3(2) - 1$, $8 = 3(3) - 1$, $11 = 4(3) - 1$, and so on, so that we might try

$$f(y) = 3(y + 1) - 1 \quad y = 1, 2, 3, \ldots$$

However, upon algebraic simplification, we find that

$$3(y + 1) - 1 = 3y + 3 - 1 = \text{_____}$$

as in 1.

3. We see, then, that if two students find two seemingly different functions using position as the independent variable and these two functions generate the same sequence, algebraic manipulation will show that the two functions are identical.

Statisticians generally work with sequences of measurements and therefore need a convenient way of referring to the first, second, . . . , or last measurement observed. Suppose that a statistician were working with measurements that were the construction costs per mile for ten different sections of an interstate highway. Wishing to use the variable c to designate cost, he would probably write c_1 (c-sub-one) to designate the cost for the first section, c_2 to designate the cost for the second section, and so on. Hence, he would refer to his measurements as

$$c_1, c_2, c_3, \ldots, c_9, c_{10}.$$

In this case we would write a typical element of this sequence as

$$f(i) = \underline{\hspace{2cm}} \qquad i = 1, 2, \ldots, 10,$$

c_i

using i as the position variable. Hence, $f(3) = \underline{\hspace{2cm}}$, the cost of the third section.

c_3

2.4 Summation Notation

Consider the expression

$$\sum_{y=2}^{4} f(y).$$

1. The Greek letter Σ (upper case sigma) is an instruction to perform the operation of $\underline{\hspace{2cm}}$.

addition

2. $f(y)$ is the yth term of the sequence $\{f(1), f(2), f(3), \ldots\}$ and is called the typical $\underline{\hspace{2cm}}$ of summation.

element

3. The notation "$y = 2$" found below the symbol Σ indicates two things:
   a. The letter y is to be used as the position variable or the *variable of summation.*
   b. The first term in the sum is to be $f(2)$, the $\underline{\hspace{2cm}}$ term in the sequence.

second

   The number of the first term in the sum is usually referred to as the *lower limit of summation.* In this problem, the lower limit of summation is $\underline{\hspace{2cm}}$.

2

4. The number "4" above the symbol Σ indicates that the last term in the sum is to be $f(4)$. In general this number is called the *upper limit of summation.*

5. The total expression $\displaystyle\sum_{y=2}^{4} f(y)$ is the instruction to add the second, third, and fourth terms of the sequence $\{f(1), f(2), \ldots\}$

Example 2.6

Evaluate $\displaystyle\sum_{y=1}^{4} y^2$.

Solution

$$\sum_{y=1}^{4} y^2 = 1^2 + 2^2 + 3^2 + 4^2$$

$$= 1 + 4 + 9 + 16$$

30

$$= \underline{\hspace{2cm}}.$$

Example 2.7

Evaluate $\displaystyle\sum_{y=2}^{4} (y + 2)$.

Solution

6; 15

$$\sum_{y=2}^{4} (y + 2) = 4 + 5 + \underline{\hspace{2cm}} = \underline{\hspace{2cm}}.$$

One must pay close attention to parentheses in problems involving summations. Notice the difference between the next two problems:

11

1. $\displaystyle\sum_{x=1}^{3} (x^2 - 1) = (1^2 - 1) + (2^2 - 1) + (3^2 - 1) = \underline{\hspace{2cm}}.$

13

2. $\displaystyle\sum_{x=1}^{3} x^2 - 1 = (1^2 + 2^2 + 3^2) - 1 = \underline{\hspace{2cm}}.$

In the first problem, the typical element of summation is $(x^2 - 1)$, which means that within each term one is subtracted from x^2. In the second problem, the typical element of summation is $\underline{\hspace{2cm}}$, so that after summing,

x^2

one is subtracted from $\displaystyle\sum_{x=1}^{3} x^2$. The placement or absence of a set of parentheses is crucial in defining the typical element of summation.

Example 2.8
Evaluate the following summations:

74

a. $\displaystyle\sum_{x=3}^{5} (x^2 + 2x) = 15 + 24 + 35 = \underline{\hspace{2cm}}.$

12

b. $\displaystyle\sum_{z=1}^{3} z/3 + 10 = (1/3 + 2/3 + 3/3) + 10 = \underline{\hspace{2cm}}.$

c. $\displaystyle\sum_{i=1}^{4} (i^2 - i + 1) = 1 + 3 + 7 + 13 = \underline{\hspace{2cm}}.$ 24

d. $\displaystyle\sum_{j=1}^{4} (j^2 - j) + 1 = (0 + 2 + 6 + 12) + 1 = \underline{\hspace{2cm}}.$ 21

Notice that any symbol can be used as a valid variable of summation. In fact the letters i and j are very commonly used as position variables in statistical problems.

Example 2.9
The following measurements represent the cost in cents of five different 7 oz. cans of aerosol deodorant:

$$x_1 = 119, x_2 = 98, x_3 = 79, x_4 = 89, x_5 = 95.$$

Using these values, evaluate the following summations:

a. $\displaystyle\sum_{i=1}^{5} x_i.$ d. $\displaystyle\sum_{i=1}^{5} (x_i - 96)^2.$

b. $\displaystyle\sum_{i=1}^{5} x_i/5.$ e. $\displaystyle\sum_{i=1}^{5} x_i^2 - \left(\sum_{i=1}^{5} x_i\right)^2 \Big/ 5.$

c. $\displaystyle\sum_{i=1}^{5} (x_i - 96).$

Solution
a. In each of these five problems, the letter i is used as a position variable to designate the ith measurement (cost) in the group of five measurements. Therefore

$$\sum_{i=1}^{5} x_i = x_1 + x_2 + x_3 + x_4 + x_5$$

$$= 119 + 98 + 79 + 89 + 95$$

$$= \underline{\hspace{2cm}}.$$ 480

b. From part a., $\displaystyle\sum_{i=1}^{5} x_i = 480$, so that

96

$$\sum_{i=1}^{5} x_i/5 = 480/5 = \underline{\hspace{2cm}}.$$

c. This problem can be solved directly as

$$\sum_{i=1}^{5} (x_i - 96) = (119 - 96) + (98 - 96) + (79 - 96) + (89 - 96)$$

$$+ (95 - 96)$$

7; 1

$$= 23 + 2 - 17 - \underline{\hspace{1.5cm}} - \underline{\hspace{1.5cm}}$$

0

$$= \underline{\hspace{2cm}}.$$

d. Using the results of part c.,

$$\sum_{i=1}^{5} (x_i - 96)^2 = (23)^2 + (2)^2 + (-17)^2 + (-7)^2 + (-1)^2$$

872

$$= \underline{\hspace{2cm}}.$$

e. To compute this summation, we first need to find

$$\sum_{i=1}^{5} x_i^2 = 119^2 + 98^2 + 79^2 + 89^2 + 95^2$$

$$= 46952.$$

480

$$\sum_{i=1}^{5} x_i^2 - \left(\sum_{i=1}^{5} x_i\right)^2 / 5 = 46952 - (\underline{\hspace{2cm}})^2/5$$

46080

$$= 46952 - \underline{\hspace{2cm}}$$

872

$$= \underline{\hspace{2cm}}.$$

(It is no accident that the answers to d. and e. are identical.)

Self-Correcting Exercises 2B

1. $h(y) = 2(1/3)^y$. Find $\displaystyle\sum_{y=1}^{3} h(y)$.

2. $g(x) = 2x^2 - 5$. Find $\displaystyle\sum_{x=1}^{4} g(x)$.

Use the following set of measurements to answer exercises 3 through 5.

i	1	2	3	4	5	6	7	8	9	10
x_i	-1	2	1	0	4	-3	1	6	-5	-2

3. $\displaystyle\sum_{i=1}^{10} x_i .$

4. $\displaystyle\sum_{i=1}^{10} x_i^2 .$

5. $\displaystyle\sum_{i=1}^{10} x_i / 10.$

2.5 Summation Theorems

We will now review three theorems involving summations. These theorems will prove useful when evaluating statistical descriptive measures that characterize a distribution of measurements.

$$\text{Theorem 2.1} \qquad \sum_{x=1}^{n} c = nc.$$

We say that c is a constant because c does not depend upon the variable of summation. One may encounter sums with a lower limit different from one. A useful device for arriving at the correct answer in this case is illustrated by the following example:

Example 2.10

Evaluate $\displaystyle\sum_{x=4}^{10} 5.$

Solution

1. Now $\displaystyle\sum_{x=1}^{10} 5$ has _____ terms while $\displaystyle\sum_{x=1}^{3} 5$ has _____ terms.

10; 3

2. But $\displaystyle\sum_{x=4}^{10} 5 = \sum_{x=1}^{10} 5 - \sum_{x=1}^{3} 5.$

7; 35

3. $\displaystyle\sum_{x=4}^{10} 5$ has _____ terms and hence $\displaystyle\sum_{x=4}^{10} 5 = (10-3)5 =$ _____.

$$\begin{bmatrix} \textit{Theorem 2.2} & \displaystyle\sum_{x=1}^{n} cf(x) = c\sum_{x=1}^{n} f(x). \end{bmatrix}$$

An equivalent form of this theorem is $\displaystyle\sum_{i=1}^{n} cx_i = c\sum_{i=1}^{n} x_i.$

6; 150

Thus, $\displaystyle\sum_{x=1}^{3} 25x = 25\sum_{x=1}^{3} x = 25(\underline{1+2+3}) =$ _____.

$$\begin{bmatrix} \textit{Theorem 2.3} & \displaystyle\sum_{x=1}^{n} [f(x) \pm g(x)] = \sum_{x=1}^{n} f(x) \pm \sum_{x=1}^{n} g(x). \end{bmatrix}$$

Thus, $\displaystyle\sum_{x=1}^{6} (x-5) = \sum_{x=1}^{6} x - \sum_{x=1}^{6} 5$

6(5)

$$= 21 - \underline{\hspace{2cm}}$$

−9

$$= \underline{\hspace{2cm}}.$$

Theorems 2.1, 2.2, and 2.3 may be used in combination as in the following example:

Example 2.11

Evaluate $\displaystyle\sum_{x=1}^{5} (3x^2 + 5x - 2).$

Solution
1. Using Theorem 2.3,

$$\sum_{x=1}^{5} (3x^2 + 5x - 2) = \sum_{x=1}^{5} 3x^2 + \sum_{x=1}^{5} 5x - \sum_{x=1}^{5} 2.$$

2. The first two terms can be simplified using Theorem 2.2 to give

$$\sum_{x=1}^{5} (3x^2 + 5x - 2) = 3 \sum_{x=1}^{5} x^2 + 5 \sum_{x=1}^{5} x - \sum_{x=1}^{5} 2.$$

3. Finally, using Theorem 2.1 to simplify $\sum_{x=1}^{5} 2 = 5(2)$,

$$\sum_{x=1}^{5} (3x^2 + 5x - 2) = 3(55) + 5(15) - 10$$

$$= \underline{\hspace{2cm}}.$$ 230

Example 2.12

Evaluate $\sum_{x=1}^{5} (x - 4)^2$.

Solution

1. If the element of summation is not written as a simple sum or difference of terms, it may be possible to convert it to a sum or difference by algebraic manipulation.
2. The typical element of summation is $(x - 4)^2$. But $(x - 4)^2 = x^2 - 8x +$ \underline{\hspace{2cm}}, so that 16

$$\sum_{x=1}^{5} (x - 4)^2 = \sum_{x=1}^{5} (x^2 - 8x + 16)$$

$$= \sum_{x=1}^{5} x^2 - 8 \sum_{x=1}^{5} x + 5(16)$$

$$= 55 - 8(\underline{\hspace{1.5cm}}) + \underline{\hspace{1.5cm}}$$ 15; 80

$$= \underline{\hspace{2cm}}.$$ 15

Example 2.13

Using summation theorems, rewrite the following summation as sums or differences of terms.

$$\sum_{i=1}^{n} (x_i - c)^2.$$

Solution

1. Following the technique given in Example 2.12, the typical element of summation is $(x_i - c)^2$, which when algebraically squared equals $(x_i^2 - 2cx_i + c^2)$.

2. Using the summation theorems, write

nc^2

$$\sum_{i=1}^{n} (x_i - c)^2 = \sum_{i=1}^{n} x_i^2 - 2c \sum_{i=1}^{n} x_i + \underline{\hspace{2cm}}.$$

3. Note that until the values of $x_1, x_2, \ldots, x_n$ and c are given, we cannot proceed any further with this problem.

Example 2.14

Use the results of Example 2.13 to find $\sum_{i=1}^{5} (x_i - 3)^2$ if

$$x_1 = 2, x_2 = 4, x_3 = 1, x_4 = 3, x_5 = 4.$$

Solution

1. From Example 2.13,

$$\sum_{i=1}^{5} (x_i - 3)^2 = \sum_{i=1}^{5} x_i^2 - 6 \sum_{i=1}^{5} x_i + 5(9).$$

Therefore, we need to find $\sum_{i=1}^{5} x_i^2$ and $\sum_{i=1}^{5} x_i$.

46

2.
$$\sum_{i=1}^{5} x_i^2 = 2^2 + 4^2 + 1^2 + 3^2 + 4^2 = \underline{\hspace{2cm}}.$$

14

$$\sum_{i=1}^{5} x_i = 2 + 4 + 1 + 3 + 4 = \underline{\hspace{2cm}}.$$

3. Collecting results we have

46; 14

$$\sum_{i=1}^{5} (x_i - 3)^2 = \underline{\hspace{2cm}} - 6(\underline{\hspace{2cm}}) + 45$$

7

$$= \underline{\hspace{2cm}}.$$

Although this method seems slow and sluggish and you would rather use

$\sum\limits_{i=1}^{5} (x_i - 3)^2$ directly for this problem, when the number of observations, n, is larger than 5 and the measurements are not small whole numbers, the expanded form of the summation will prove to be the simpler of the two to use in practice.

Self-Correcting Exercises 2C

1. Evaluate $\sum\limits_{x=1}^{10} 3$.

2. Evaluate $\sum\limits_{i=7}^{14} 4$.

3. Refer to Self-Correcting Exercises 2B, problems 3, 4, and 5. Using the set of measurements given, evaluate the following summations:

   a. $\sum\limits_{i=3}^{8} (x_i - 4)$.

   b. $\sum\limits_{i=1}^{10} (x_i - 5)^2$.

   c. $\sum\limits_{i=1}^{5} (x_i^2 - x_i)$.

   d. $\sum\limits_{i=1}^{10} x_i^2 - \left(\sum\limits_{i=1}^{10} x_i\right)^2 \Big/ 10$.

EXERCISES

1. $f(y) = 3y - 2u$. Find $f(u)$.
2. $G(u) = 3u^3 - 2u$. Find $G(a)$.
3. $h(y) = 2(1/3)^y$. Find $h(0)$ and $h(2)$.
4. $f(u) = 2u^2 - 3u + 5$. Find $f(1/x)$.
5. $F(u) = 2u^2 - 3/u$ and $g(v) = 1/v$.
   Find $F[g(v)]$.

6. Expand $\displaystyle\sum_{i=3}^{6} (x_i - a)$.

7. Express the following in a more compact form by using summation notation:

$$(x_3 - m)^2 + (x_4 - m)^2 + (x_5 - m)^2 + (x_6 - m)^2.$$

8. $\displaystyle\sum_{u=1}^{3} (3x + u!) = $ _____.

9. $\displaystyle\sum_{x=0}^{3} \frac{(.75)}{x! \, (3 - x)!} = $ _____.

10. Evaluate $\displaystyle\sum_{x=7}^{8} 4x$.

11. If $\displaystyle\sum_{i=1}^{25} x_i = 50$, find $\displaystyle\sum_{i=1}^{25} (2x_i - 3)$.

12. If $\displaystyle\sum_{i=1}^{25} x_i = 50$ and $\displaystyle\sum_{i=1}^{25} x_i^2 = 250$,

find $\displaystyle\sum_{i=1}^{25} (x_i - 10)^2$.

Chapter 3

DESCRIBING DISTRIBUTIONS OF MEASUREMENTS

3.1 Introduction

Having reviewed some essential mathematics, we return to the objective of statistics, which is making _____ about a _____ from information contained in a _____ . An obvious but often ignored requirement is that the sample be drawn from the population of interest to the experimenter.

inferences; population sample

   How are inferences made? First, we must be able to describe data in a straightforward pictorial or graphical form. Second, we must be able to reconstruct this visual representation using numerical descriptive measures that describe the salient characteristics of the visual representation. For example, where is the middle of the distribution? Are the measurements tightly grouped or widely scattered? Whether the set of data under consideration comprises an entire population, or is merely a sample from a population, we must be able to agree upon numerical measures that describe the data. Inferences about a population can then be made in terms of the population by using the relevant information contained in the sample.

   This chapter will deal with two very general methods of describing data:
1. Graphical methods.
2. Numerical methods.

3.2 Frequency Distributions

Graphical methods attempt to present the set of measurements in pictorial form so as to give the reader an adequate visual description of the measurements. Let us discuss a graphical method by examining the following data.

Example 3.1

The following data are the numbers of correct responses on a recognition test consisting of 30 items, recorded for 25 students:

25	29	23	27	25
23	22	25	22	28
28	24	17	24	30
19	17	23	21	24
15	20	26	19	23

30

15

1. First find the highest score, which is _____ , and the lowest score, which is _____ . These two scores indicate that the measurements have a range of 15.
2. To determine how the scores are distributed between 15 and 30, we divide this interval into subintervals of equal length. The interval from 15 to 30 could be divided into from 5 to 20 subintervals, depending upon the number of measurements available. Wishing to obtain about 7 subintervals, a suitable width is determined by dividing 30 − 15 = 15 by 7. The integer

2

_____ would seem to provide a most satisfactory subinterval width for these data.

28.5; 30.5

3. Utilizing the subinterval boundary points, 14.5, 16.5, 18.5, 20.5, 22.5, 24.5, 26.5, _____ , and _____ , we guarantee that none of the given measurements will fall on a boundary point. Thus, each measurement falls into only one of the subintervals or classes.
4. We now proceed to tally the given measurements and record the class frequencies in a table. Fill in the missing information.

Tabulation of Data for Histogram

Class (i)	Class Boundary	Tally	Frequency (f_i)	Relative Frequency (f_i/n)
1	14.5–16.5	1	1	1/25
2	16.5–18.5	11	2	2/25
3	18.5–20.5	_____	3	3/25
4	20.5–22.5	111	3	_____
5	22.5–24.5	⊮⊩ 11	_____	7/25
6	_____	1111	4	4/25
7	26.5–28.5	_____	3	3/25
8	28.5–30.5	11	2	2/25

111
3/25
7
24.5-26.5
111

5. The number of measurements falling in the ith class is called the ith class frequency and is designated by the symbol f_i. Of the total number of measurements, the fraction falling in the ith class is called the _____ frequency in the ith class. Given n measurements, the relative frequency in the ith class is given as f_i/n. As a check on your tabulation, remember that for k classes,

relative

n

a. $$\sum_{i=1}^{k} f_i = \underline{\hspace{2cm}}$$

and

1

b. $$\sum_{i=1}^{k} f_i/n = \underline{\hspace{2cm}} .$$

6. With the data so tabulated, we can now use a frequency histogram (plotting frequency against classes) or a relative frequency histogram (plotting relative frequency against classes) to describe the data. The two histograms are identical except for scale.
   a. Study the following histogram based on our data:

   Frequency Histogram

   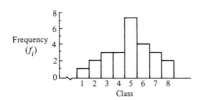

   b. Complete the following relative frequency histogram for the same data:

   Relative Frequency Histogram

   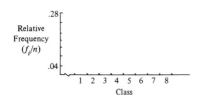

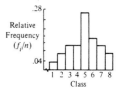

   c. When completed, the histograms in a. and b. should appear identical.
7. By examining the tabulation found in 4., answer the following questions:
   a. What fraction of the students had scores less than 20.5? _____ .

   6/25 or 24%

   b. What fraction of the students had scores greater than 26.5?

   _____ .

   5/25 or 20%

   c. What fraction of the students had scores between 20.5 and 26.5?

   _____ .

   14/25 or 56%

8. As the number of measurements in the sample increases, the sample histogram should resemble the population histogram more and more. Thus, to estimate the fraction of students in the entire population that would have scores greater than 26.5, we could use our sample histogram, estimating this fraction to be _____ or _____ .

   5/25; 20%

9. A relative frequency histogram is often called a relative frequency distribution because it displays the manner in which the data are distributed along the horizontal axis of the graph. The rectangular bars above the class intervals in the relative frequency histogram can be given two interpretations.
   a. The height of the bar above the ith class would represent the fraction of observations falling in the ith class.
   b. The height of the bar above the ith class would also represent the probability that a measurement drawn at random from this sample will belong to the ith class.

10. Complete the following statements based on the data tabulation in 4.

a. The probability that a measurement drawn at random from this data will fall in the interval 22.5 to 24.5 is _____.

7/25

b. The probability that a measurement drawn at random from this data will be greater than 18.5 is _____.

22/25

c. The probability that a measurement drawn at random from this data will be less than 24.5 is _____.

16/25

Example 3.2

The following data represent the burning times for an experimental lot of fuses measured to the nearest tenth of a second:

5.2	3.8	5.7	3.9	3.7
4.2	4.1	4.3	4.7	4.3
3.1	2.5	3.0	4.4	4.8
3.6	3.9	4.8	5.3	4.2
4.7	3.3	4.2	3.8	5.4

Construct a relative frequency histogram for these data.

Solution

Fill in the missing entries in the table below.

Tabulation of Data

Class	Class Boundary	Tally	Frequency	Relative Frequency
1	2.45–2.95	1	1	.04
2	2.95–3.45	111	3	_____
3	3.45–3.95	TТᏧ 1	_____	.24
4	3.95–4.45	_____	7	.28
5	4.45–4.95	1111	4	_____
6	_____	111	3	.12
7	5.45–5.95	1	1	.04

.12
6
TᏧᏧ 11
.16
4.95–5.45

Relative Frequency Histogram

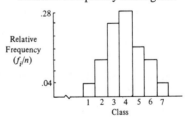

Complete the following statements based on the preceding tabulation.

a. The probability that a measurement drawn at random from this sample is greater than 4.45 is _____.

8/25

b. The probability that a measurement drawn at random from this sample is less than 3.45 is _____ .

c. An estimate of the probability that a measurement drawn at random from the sampled population would be in the interval 3.45 to 4.45 is _____ .

4/25

13/25

Self-Correcting Exercises 3A

1. The following data are the ages in years of 42 students enrolled in an adult education class:

51	32	31	33	23	52
23	21	55	34	38	32
49	35	26	29	50	34
30	19	41	39	41	27
25	21	18	36	35	28
44	44	59	28	23	46
27	37	42	32	43	30

a. Find the range of these data.

b. Using about 10 intervals of equal width, set up class boundaries to be used in the construction of a frequency distribution and complete the tabulation of the data.

c. Construct a frequency histogram for these data.

d. For these same data, construct a frequency histogram utilizing about 6 intervals.

e. Which histogram presents the salient points of these data more clearly?

2. The following are the annual rates of profit on stockholders' equity after taxes in percentage for 32 industries:

10.6	10.8	14.8	10.8
12.5	6.0	10.7	11.0
14.6	6.0	12.8	10.1
7.9	5.9	10.0	10.6
10.8	16.2	18.4	10.7
10.6	13.3	8.7	15.4
6.5	10.1	8.7	7.5
11.9	9.0	12.0	9.1

a. Present a relative frequency histogram for these data utilizing 7 intervals of length 2, beginning at 5.55.

Using your histogram (or tabulation) answer the following questions:

b. What is the probability that an industry drawn at random from this distribution has a rate of profit greater than 15.55%?

c. What is the probability that an industry drawn at random has a rate of profit less than 9.55?

d. What is the probability that an industry drawn at random has a rate of profit greater than 9.55 but less than 15.55?

3.3 Numerical Descriptive Measures

The chief advantage to using a graphical method is its visual representation of the data. Many times, however, we are restricted to reporting our data verbally. In this case a graphical method of description cannot be used. The greatest disadvantage to a graphical method of describing data is its unsuitability for making inferences, since it is difficult to give a measure of goodness for a graphical inference. Therefore, we turn to *numerical descriptive measures*. We seek a set of numbers that characterizes the frequency distribution of the measurements and at the same time will be useful in making inferences.

3.4 Numerical Measures of Central Tendency

We first consider two of the more important measures of central tendency that attempt to locate the center of the frequency distribution.

> The *mean* of a set of measurements, $y_1, y_2, \ldots, y_n$, is defined to be the sum of the measurements divided by n. The symbol $\bar{y}$ is used to designate the sample mean while the Greek letter μ is used to designate the population mean.

The sample mean can be shown to have very desirable properties as an inference maker. In fact, we will use $\bar{y}$ to estimate the population mean, μ. Using summation notation, we can define the sample mean by formula as

$$\sum_{i=1}^{n} y_i/n \qquad \bar{y} = \underline{\hspace{3cm}}.$$

Example 3.3
Find the mean of the following measurements:

$$2, \quad 5, \quad 7, \quad 10, \quad 11, \quad 13.$$

Solution

48

$$\sum_{i=1}^{6} y_i = \underline{\hspace{2cm}}$$

48

$$\bar{y} = \frac{\Sigma y_i}{n} = \frac{\underline{\hspace{1.5cm}}}{6}$$

8

$$\bar{y} = \underline{\hspace{2cm}}.$$

In addition to being an easily calculated measure of central tendency, the mean is also easily understood by all users. The calculation of the mean

utilizes all of the measurements and can always be found exactly. One disadvantage of using the mean to measure central tendency is well known to
any student who has had to pull up one low test score: the mean (is, is not)
greatly affected by extreme values.

    One might be unwilling to accept the average property value of $75,000 for
a given area as an acceptable measure of the middle property value if it were
known that (a) the property value of a residence owned by a millionaire was
included in the calculation, and (b) excluding this residence, the property
values ranged from $15,000 to $30,000. A more realistic measure of central
tendency in this situation might be the property value such that 50% of the
property values are less than this value and 50% are greater.

> The *median* of a set of n measurements, $y_1, y_2, \ldots, y_n$, is the
> value of y that falls in the middle when the measurements are
> arranged in order of magnitude. When n is even and there are two
> middle values, the median is taken to be the simple average of the
> two middle values.

Example 3.4
Find the median of the following set of measurements:

    5, 3, 2, 7, 4.

Solution
1. Arranging the measurements in order of magnitude, we have

    2, 3, 4, 5, 7.

2. The median will be the _____ ordered value. Hence the median
   is _____ .

Example 3.5
Find the median of the following set of measurements:

    10, 8, 13, 14, 9, 8.

Solution
1. Arranging the measurements in order of magnitude, we have

    8, 8, 9, 10, 13, 14.

2. Since $n = 6$ is even, the median will be the average of the _____
   and _____ ordered values. Hence

$$\text{median} = \frac{\_\_\_\_\_ + \_\_\_\_\_}{2}$$

$$= \_\_\_\_\_ .$$

is	
third	
4	
third	
fourth	
9; 10	
9.5	

Example 3.6

Find the mean and median of the following data:

$$5, \ 7, \ 8, \ 10, \ 10, \ 11, \ 13, \ 14.$$

Solution

78; 78; 9.75

1. $\displaystyle\sum_{i=1}^{8} y_i = \underline{\hspace{1.5cm}}; \quad \bar{y} = \dfrac{\Sigma y_i}{n} = \dfrac{}{8} = \underline{\hspace{1.5cm}}.$

2. To find the median, we note that the measurements are already arranged in order of magnitude and that $n = 8$ is even. Therefore the median will be the average of the fourth and fifth ordered values.

10

$$\text{median} = \frac{10 + 10}{2} = \underline{\hspace{1.5cm}}.$$

In the last example, the mean and median gave reasonably close numerical values as measures of central tendency. However, if the measurement $y_9 = 30$ were added to the eight measurements given, the recalculated mean would be

12

$\bar{y} = \underline{\hspace{1.5cm}}$, but the median would remain at 10.

3.5 Measures of Variability

Having found measures of central tendency, we next consider measures of the variability or dispersion of the data. A measure of variability is necessary since a measure of central tendency alone does not adequately describe the data. Consider the two sets of data which follow:

Set I. $x_1 = 9$	Set II. $y_1 = 1$

10; 10

$x_2 = 10 \qquad \bar{x} = \underline{\hspace{1cm}}$

$y_2 = 10 \qquad \bar{y} = \underline{\hspace{1cm}}$

$x_3 = 11$

$y_3 = 19$

10

Both of these sets of data have a mean equal to _____. However, the second set of measurements displays much more variability about the mean than does the first set.

In addition to a measure of central tendency, a measure of variability is indispensable as a descriptive measure for a set of data. A manufacturer of

little

machine parts would want very (little, much) variability in his product in order to control oversized or undersized parts, while an educational testing

large

service would be satisfied only if the test scores showed a (large, small) amount of variability in order to discriminate among people taking the examination.

We have already used the simplest measure of variability, the range.

The *range* of a set of measurements is the difference between the largest and smallest measurements.

Example 3.7
Find the range for each of the following sets of data:

Set I.

23	73	34	74
28	29	26	17
88	8	52	49
37	96	32	45
81	62	23	62

Range = 96 – _____ = _____.   8; 88

Set II.

8.8	6.7	7.1	2.9
9.0	0.2	1.2	8.6
6.3	6.4	2.1	8.8

Range = 9.0 – _____ = _____.   0.2; 8.8

By examining the following distributions, it is apparent that although the range is a simply calculated measure of variation, it alone is not adequate. Both distributions have the same range, but display different variability.

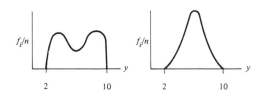

In looking for a more sensitive measure of variability, we can extend the concept of the median as follows:

> Let $y_1, y_2, \ldots, y_n$ be a set of n measurements arranged in order of increasing magnitude. The pth *percentile* is that value of y such that $p\%$ of the measurements are less than y and $(100 - p)\%$ are greater than y.

The 25th percentile is called the *lower quartile,* while the 75th percentile is called the *upper quartile.* Percentiles are more sensitive than the range in measuring variability, but have the disadvantage that several percentiles must be calculated to provide an adequate description of the data.

We base the next important measure of variability on the dispersion of the data about the sample mean, $\bar{y}$. Define the quantity $(y_i - \bar{y})$ as the ith deviation from the mean. Large deviations indicate (more, less) variability of the data than do small deviations. We could utilize these deviations in different ways.

    more

1. If we attempt to use the average of the n deviations, we find that the sum of the deviations is _____. To avoid a zero sum, we could use the average of the absolute values of the deviations. This measure, called the *mean deviation,* is difficult to calculate, and one cannot easily give a measure of its goodness as an inference maker.

    zero

2. More efficient use of the data is achieved by averaging the sum of squares

of these deviations. This measure, called the *sample variance,* is given by

$$s'^2 = \frac{\sum\limits_{i=1}^{n} (y_i - \bar{y})^2}{n} .$$

large; small

Large values of s'^2 indicate (large, small) variability, while (large, small) values indicate small variability. To differentiate between the sample and the population variances, we use σ^2 to represent the population variance.

Since s'^2 is not in the original units of measurements, we can return to these units by defining the standard deviation.

> The *standard deviation, s',* is the positive square root of the variance. That is,

$$s' = \sqrt{s'^2} = \sqrt{\frac{\sum\limits_{i=1}^{n} (y_i - \bar{y})^2}{n}} .$$

In the same vein, the population standard deviation is given as

$$\sigma = \sqrt{\sigma^2}.$$

Example 3.8
Calculate the sample mean, variance, and standard deviation for the following data:

$$4, \ 2, \ 3, \ 5, \ 6.$$

Solution
Arrange the measurements in the following way, first finding the mean,

4

$\bar{y} = $ _____ .

y_i	$y_i - \bar{y}$	$(y_i - \bar{y})^2$
4	0	0
2	-2	4
3	-1	1
5	1	1
6	2	4

0; 10

$\Sigma y_i = 20$	$\Sigma(y_i - \bar{y}) = $ _____	$\Sigma(y_i - \bar{y})^2 = $ _____

After finding the mean, complete the second column and note that its sum is zero. The variance is

$$s'^2 = \frac{\Sigma(y_i - \bar{y})^2}{n} = \frac{\underline{\qquad}}{5} = \underline{\qquad},$$

10; 2

while the standard deviation is

$$s' = \sqrt{2} = \underline{\qquad}.$$

1.414

Note: We shall introduce a shortcut formula for calculating

$$\sum_{i=1}^{n} (y_i - \bar{y})^2;$$

more examples will be given then.

Self-Correcting Exercises 3B

1. Fifteen brands of breakfast cereal were judged by nutritionists according to four criteria: taste, texture, nutritional value, and popularity with the buying public. Each brand was rated on a 0–5 scale for each criterion and the sum of the four ratings reported. (A high score with respect to the maximum of 20 points indicates a good evaluation of the brand.)

9	8	16	17	10
15	12	6	12	13
10	13	19	11	9

Find the mean and the median scores for these data. Compare their values.

2. The number of daily arrivals of cargo vessels at a west coast port during an 11-day period are given below:

3	2	0
5	4	4
2	3	2
7	1	

a. Compare the mean and median for these data.
b. Calculate the range of the data.
c. Calculate the standard deviation of the number of arrivals per day during this 11-day period. (As an intermediate check on calculations, remember that the sum of deviations must be zero.)

3.6 On the Practical Significance of the Standard Deviation

Having defined the mean and standard deviation, we now introduce two theorems which will use both of these quantities in more fully describing a set of data.

Tchebysheff's Theorem: Given a number, k, greater than or equal to one and a set of n measurements, $y_1, y_2, \ldots, y_n$, at least $(1 - 1/k^2)$ of the measurements will lie within k standard deviations of their mean.

The importance of this theorem is due to the fact that it applies to any set of measurements. It applies to a population using the population mean, μ, and the population standard deviation, σ, and it applies to a sample from a given population using $\bar{y}$ and s', the sample mean and sample standard deviation. Since this theorem applies to any set of measurements it is of necessity a

conservative

_____ theorem. It is therefore very important to stipulate that

at least

_____ _____ $(1 - 1/k^2)$ of the measurements will lie within k standard deviations of their mean.

Complete the following chart for the values of k given:

k	Interval $\bar{y} \pm ks'$	Interval Contains at Least the Fraction $(1 - 1/k^2)$
1	$\bar{y} \pm s'$	_____
2	$\bar{y} \pm 2s'$	_____
3	$\bar{y} \pm 3s'$	_____
10	$\bar{y} \pm 10s'$	_____

0
3/4
8/9
99/100

Example 3.9

The mean and variance of a set of $n = 20$ measurements are 35 and 25, respectively. Use Tchebysheff's Theorem to describe the distribution of these measurements.

Solution

Collecting pertinent information we have:

$$\bar{y} = 35$$
$$s'^2 = 25$$
$$s' = \sqrt{25} = 5.$$

25
45
20
50

15; 55

1. At least 3/4 of the measurements lie in the interval $35 \pm 2(5)$ or _____ to _____.
2. At least 8/9 of the measurements lie in the interval $35 \pm 3(5)$ or _____ to _____.
3. At least 15/16 of the measurements lie in the interval $35 \pm 4(5)$ or _____ to _____.

Example 3.10

If the mean and variance of a set of $n = 50$ measurements are 42 and 36, respectively, describe these measurements using Tchebysheff's Theorem.

Solution

Pertinent information: $\bar{y} = 42 \quad s'^2 = 36 \quad s' = 6.$

1. At least 3/4 of the measurements lie in the interval $42 \pm 2(6)$ or _____
   to _____.

2. At least 8/9 of the measurements lie in the interval $42 \pm 3(6)$ or _____
   to _____.

3. At least 15/16 of the measurements lie in the interval $42 \pm 4(6)$ or
   _____ to _____.

	30
	54
	24
	60
	18; 66

Empirical Rule: Given a distribution of measurements which is approximately bell-shaped, the interval
a. $\mu \pm \sigma$ contains approximately 68% of the measurements.
b. $\mu \pm 2\sigma$ contains approximately 95% of the measurements.
c. $\mu \pm 3\sigma$ contains approximately 99.7% of the measurements.

This rule holds reasonably well for any set of measurements that possesses a distribution that is mound-shaped. Bell-shaped or mound-shaped is taken to mean that the distribution has the properties associated with the normal distribution whose graph is given in your text and elsewhere in this study guide.

Example 3.11
A random sample of 100 oranges was taken from a grove and individual weights measured. The mean and variance of these measurements were 7.8 ounces and 0.36 (ounces)^2, respectively. Assuming the measurements produced a mound-shaped distribution, describe these measurements using the Empirical Rule.

Solution
First find the intervals needed.

k	$\bar{y} \pm ks'$	$\bar{y} - ks'$	to	$\bar{y} + ks'$	
1	$\bar{y} \pm s'$	_____	to	_____	7.2; 8.4
2	$\bar{y} \pm 2s'$	_____	to	_____	6.6; 9.0
3	$\bar{y} \pm 3s'$	_____	to	_____	6.0; 9.6

Then approximately

a. _____% of the measurements lie in the interval _____ to
   _____,

b. _____% of the measurements lie in the interval _____ to
   _____,

c. _____% of the measurements lie in the interval _____ to
   _____.

68; 7.2
8.4
95; 6.6
9.0
99.7; 6.0
9.6

When n is small, the distribution of measurements (would, would not) be mound-shaped and as such the Empirical Rule (would, would not) be appropriate in describing this data. Since Tchebysheff's Theorem applies to any set of measurements, it can be used regardless of the size of n.

would not
would not

3.7   A Short Method for Calculating the Variance

The calculation of $s'^2 = \dfrac{\Sigma (y_i - \bar{y})^2}{n}$ requires the calculation of the quan-

tity $\displaystyle\sum_{i=1}^{n} (y_i - \bar{y})^2$. To facilitate this calculation, we introduce the identity

$$\left[\sum_{i=1}^{n} (y_i - \bar{y})^2 = \sum_{i=1}^{n} y_i^2 - \frac{\left(\displaystyle\sum_{i=1}^{n} y_i \right)^2}{n} \right]$$

the proof of which is given in your text. This computation requires

1. The ordinary arithmetic sum of the measurements, $\displaystyle\sum_{i=1}^{n} y_i$.

2. The sum of the squares of the measurements, $\displaystyle\sum_{i=1}^{n} y_i^2$.

Note the distinction between $\displaystyle\sum_{i=1}^{n} y_i^2$ and $\left(\displaystyle\sum_{i=1}^{n} y_i \right)^2$ used in the identity given above.

1. To calculate $\displaystyle\sum_{i=1}^{n} y_i^2$, we *first square* each measurement and *then sum* these squares.

2. To calculate $\left(\displaystyle\sum_{i=1}^{n} y_i \right)^2$, we *first sum* the measurements and *then square* this sum.

Example 3.12
Calculate s'^2 for Example 3.8.

Solution
Display the data in the following way, finding Σy_i and Σy_i^2:

y_i	y_i^2	
4	16	
2	4	
3	9	
5	25	
6	36	
$\Sigma y_i =$ _____	$\Sigma y_i^2 =$ _____	20; 90

1. We first calculate

$$\Sigma(y_i - \bar{y})^2 = \Sigma y_i^2 - \frac{(\Sigma y_i)^2}{n}$$

$$= 90 - \frac{(20)^2}{5}$$

$$= 90 - \underline{\hspace{2cm}} \qquad\qquad 80$$

$$= \underline{\hspace{2cm}}. \qquad\qquad 10$$

2. Then

$$s'^2 = \frac{\Sigma(y_i - \bar{y})^2}{n} = \frac{\underline{\hspace{1.5cm}}}{5} = \underline{\hspace{2cm}}. \qquad\qquad 10; 2$$

Example 3.13
Calculate the mean and variance of the following data: 5, 6, 7, 5, 2, 3.

Solution
Display the data in tabled form:

y_i	y_i^2	
5	25	
6	36	
7	49	
5	25	
2	4	
3	9	
$\Sigma y_i =$ _____	$\Sigma y_i^2 =$ _____	28; 148

1. $\quad \bar{y} = \dfrac{\Sigma y_i}{n} = \dfrac{28}{6} = \underline{\hspace{2cm}}.$ 4.67

2. $\quad \Sigma(y_i - \bar{y})^2 = \Sigma y_i^2 - \dfrac{(\Sigma y_i)^2}{n}$

$$= 148 - \frac{(28)^2}{6}$$

$$= 148 - 130.67$$

17.33

$$= \underline{\hspace{3cm}}.$$

17.33; 2.89

3. $\quad s'^2 = \dfrac{\Sigma(y_i - \bar{y})^2}{n} = \dfrac{\underline{\hspace{2cm}}}{6} = \underline{\hspace{2cm}}.$

Self-Correcting Exercises 3C

1. Using the data given in Exercise 2, Self-Correcting Exercises 3B, calculate the variance utilizing the shortcut formula to calculate the required sum of squared deviations. Verify that the values of the variance (and hence the standard deviation) found using both calculational forms are identical.
2. If a person were concerned about accuracy due to rounding of numbers at various stages in computation, which formula for calculating $\Sigma(y_i - \bar{y})^2$ would be preferred:

    a. $\Sigma(y_i - \bar{y})^2$ or   b. $\Sigma y_i^2 - \dfrac{(\Sigma y_i)^2}{n}$?

Defend your choice of either a. or b.

3.8 Estimating the Population Variance

Since our objective is to make inferences about the population based on sample data, it is appropriate to ask if the sample mean and variance are good estimators of their population counterparts, μ and σ^2. The fact is that $\bar{y}$ is a good estimator of μ, but s'^2 appears to (underestimate, overestimate) the population variance, σ^2, when the sample size is small.

underestimate

    The problem of underestimating σ^2 can be solved by dividing the sum of squares of deviations by $n - 1$ rather than n. We then define

$$\left[s^2 = \frac{\Sigma(y_i - \bar{y})^2}{n - 1} \right]$$

as the sample estimator of the population variance. For the most part, our objective involves making inferences about populations, rather than merely describing the sample measurements. Since "sample estimator of the population variance" is too cumbersome a name for s^2 (although correct), the quantity

$$s^2 = \frac{\Sigma(y_i - \bar{y})^2}{n - 1}$$

is usually referred to as the sample variance, and s^2 rather than s'^2 is *calculated* and *used* as the sample variance. Notice that for large samples (large n) the actual difference between s^2 and s'^2 will be very small. In any case, s'^2 will always be (smaller, larger) than s^2.

smaller

Example 3.14
Calculate s^2 and s for the data of Example 3.8.

Solution
From Example 3.8, we have $\Sigma(y_i - \bar{y})^2 = 10$, so that

$$s^2 = \frac{\Sigma(y_i - \bar{y})^2}{n - 1} = \frac{10}{4} = \underline{\hspace{2cm}} \quad \text{and } s = \sqrt{\underline{\hspace{2cm}}}$$

2.5; 2.5

$$= \underline{\hspace{2cm}}.$$

1.58

Example 3.15
Calculate $\bar{y}$ and s^2 for the measurements: 10, 12, 11, 10, 9.

Solution
Display the data in tabular form as follows:

y_i	y_i^2
10	100
12	144
11	121
10	100
9	81
$\Sigma y_i = 52$	$\Sigma y_i^2 = 546$

1. $\bar{y} = \dfrac{\Sigma y_i}{n} = \dfrac{}{5} = \underline{\hspace{2cm}}.$

52; 10.4

2. The sum of squares of deviations is calculated as

$$\Sigma(y_i - \bar{y})^2 = \Sigma y_i^2 - \frac{(\Sigma y_i)^2}{n}$$

$$= \underline{\hspace{2cm}} - \frac{(52)^2}{5}$$

546

$$= \underline{\hspace{2cm}}.$$

5.2

3. The "sample variance" is given as

$$s^2 = \frac{\Sigma(y_i - \bar{y})^2}{n - 1} = \frac{5.2}{\underline{\hspace{1cm}}} = \underline{\hspace{2cm}}.$$

4; 1.3

4. The standard deviation is given as

$$s = \sqrt{\underline{\hspace{2cm}}} = \underline{\hspace{2cm}}.$$

1.3; 1.1

Self-Correcting Exercises 3D

1. Calculate the "sample variance," s^2, for the data of Exercise 2, Self-Correcting Exercises 3B. Verify that s^2 is larger than s'^2.
2. The following measurements represent the times required for rats to run a maze correctly: 5.2, 4.2, 3.1, 3.6, 4.7, 4.8, 4.1. Calculate the sample variance and standard deviation.
3. The heights in inches of 5 men consecutively entering a doctor's office were 70, 74, 69, 71, 72. Calculate the mean and variance of these heights.

3.9 A Check on the Calculation of s

For mound-shaped or approximately normal data, one can use the range to check the calculation of s, the standard deviation. According to Tchybesheff's Theorem and the Empirical Rule, at least 3/4 and more likely 95% of a set of measurements will be in the interval $\bar{y} \pm 2s$. Hence, the sample range, R, should approximately equal $4s$, so that

$$s \approx R/4.$$

This approximation requires only that the computed value be of the same order as the approximation.

Example 3.16
Check the calculated value of s for the first set of data given in Example 3.7.

Solution
88 For these data, the range is $96 - 8 =$ _____ , and

22 $$s \approx 88/4 = \text{_____}.$$

would not Comparing 22 with the calculated value, 25.46, we (would, would not) have reason to doubt the accuracy of the calculated value.
In referring to the second set of data in Example 3.7 that consists of 12 measurements, we find that the range is $9.0 - 0.2 = 8.8$. Hence an approximation to s using $R \approx 4s$ yields

2.2 $$s \approx (8.8)/4 = \text{_____}.$$

The calculated value of s is 3.21, and this approximation is not as close as the approximation for the first set of data.
large Since extreme measurements are more likely to be observed in (large, small) samples, we can adjust the approximation to s by dividing the range by a divisor that depends upon the sample size, n. A rule of thumb to use in approximating s by using the range is given in the following table:

n	Divide Range by
5	2.5
10	3
25	4
100	5

Example 3.17
Use the range approximation to check the calculated value of s, 3.21, for the second set of data in Example 3.7.

Solution
We know that $R = 8.8$; hence for $n = 12$ measurements, we use the approximation

$$s \approx R/3 = (8.8)/3 = \underline{\hspace{1.5cm}}$$ 2.93

which more closely approximates the calculated value of s, 3.21, than did the earlier approximation, 2.2.

Example 3.18
Use the range approximation to check the calculation of s for the data given in Example 3.8.

Solution
For the five measurements, 4, 2, 3, 5, 6, the range is $6 - 2 = \underline{\hspace{1.5cm}}$. 4
Therefore, an approximation to s would be

$$s \approx R/(2.5) = 4/(2.5) = \underline{\hspace{1.5cm}}$$ 1.6

which closely agrees with the calculated value of 1.58.

3.10 The Effect of Coding on $\bar{y}$ and s^2

To simplify the calculation of $\bar{y}$ and s^2, data are frequently coded by subtracting (or adding) a constant from each measurement and/or multiplying (or dividing) each measurement by a constant. The following theorems will define the relationship between the mean and variance of the coded data and the mean and variance of the original data.

> *Theorem 3.1* Let $y_1, y_2, \ldots, y_n$ be n measurements with variance s^2 and let $x_i = y_i - c$, for $i = 1, 2, \ldots, n$. Then
>
> $$\begin{bmatrix} 1.\ \bar{x} = \bar{y} - c \ \ \text{or} \ \ \bar{y} = \bar{x} + c. \\ 2.\ s_x^2 = s^2. \end{bmatrix}$$

c — constant

Proof

$$1.\ \bar{x} = \frac{\sum\limits_{i=1}^{n} x_i}{n} = \frac{\sum\limits_{i=1}^{n} (y_i - c)}{n} = \frac{\sum\limits_{i=1}^{n} y_i}{n} - \frac{\sum\limits_{i=1}^{n} c}{n} = \bar{y} - c.$$

That is, $\bar{y} = \bar{x} + c.$

$$2.\ (n-1)s_x^2 = \sum_{i=1}^{n} (x_i - \bar{x})^2 = \sum_{i=1}^{n} (y_i - c - [\bar{y} - c])^2$$

$$= \sum_{i=1}^{n} (y_i - \bar{y})^2 = (n-1)s^2.$$

Hence, $s_x^2 = s^2.$

Example 3.19

Find the mean and variance for the 5 measurements, 13, 12, 14, 12, 10.

Solution

Let $x_i = y_i - 10$; then the coded data are 3, 2, 4, 2, 0 and

11; 33

$$\Sigma x_i = \underline{\hspace{2cm}} \qquad \Sigma x_i^2 = \underline{\hspace{2cm}}.$$

2.2

1. $\quad \bar{x} = \dfrac{\Sigma x_i}{n} = \dfrac{11}{5} = \underline{\hspace{2cm}}.$

2.2

2. $\quad s_x^2 = \dfrac{\Sigma x_i^2 - \dfrac{(\Sigma x_i)^2}{n}}{n-1} = \dfrac{33 - \dfrac{(121)}{5}}{4} = \underline{\hspace{2cm}}.$

3. Using Theorem 3.1, we find

12.2

$$\bar{y} = \bar{x} + 10 = 2.2 + 10 = \underline{\hspace{2cm}}$$

and

2.2

$$s^2 = s_x^2 = \underline{\hspace{2cm}}.$$

Theorem 3.2 Let $y_1, y_2, \ldots, y_n$ be n measurements with variance s^2 and let $x_i = ky_i$ for $i = 1, 2, \ldots, n$. Then

1. $\bar{x} = k\bar{y}$ or $\bar{y} = (1/k)\bar{x}.$
2. $s_x^2 = k^2 s^2$ or $s^2 = (1/k^2)s_x^2.$

k - constant

Proof

1. $\displaystyle \bar{x} = \frac{\sum\limits_{i=1}^{n} x_i}{n} = \frac{\sum\limits_{i=1}^{n} ky_i}{n} = \frac{k \sum\limits_{i=1}^{n} y_i}{n} = k\bar{y}$

so that $\bar{y} = (1/k)\bar{x}$.

2. $\displaystyle (n-1)s_x^2 = \sum\limits_{i=1}^{n} (x_i - \bar{x})^2 = \sum\limits_{i=1}^{n} (ky_i - k\bar{y})^2 = \sum\limits_{i=1}^{n} k^2(y_i - \bar{y})^2$

$\displaystyle = k^2 \sum\limits_{i=1}^{n} (y_i - \bar{y})^2.$

That is,

$$(n-1)s_x^2 = (n-1)k^2s^2$$

$$s_x^2 = k^2s^2$$

$$s^2 = \frac{1}{k^2}s_x^2.$$

Example 3.20

Find the mean and variance for the measurements .05, .01, .07, .03, .04, .01 using the coding $x_i = 100y_i$ for $i = 1, 2, \ldots, 6$.

Solution

The coded measurements are 5, 1, 7, 3, 4, 1 so that

$\Sigma x_i = $ _____ $\Sigma x_i^2 = $ _____ . | 21; 101

1. $\quad \bar{x} = \dfrac{21}{6} = $ _____ . | 3.5

2. $\quad s_x^2 = \dfrac{101 - \dfrac{(21)^2}{6}}{5} = $ _____ . | 5.5

3. Using Theorem 3.2, we have

$\quad \bar{y} = \dfrac{\bar{x}}{100} = \dfrac{(3.5)}{100} = $ _____ . | .035

.00055

$$s^2 = s_x^2/(100)^2 = (5.5)/10000 = \underline{\hspace{2cm}}.$$

Theorem 3.3 Let $y_1, y_2, \ldots, y_n$ be n measurements with variance s^2 and let $x_i = k(y_i - c)$ for $i = 1, 2, \ldots n$. Then

$$
\begin{bmatrix}
1.\ \bar{x} = k(\bar{y} - c) \quad \text{or} \quad \bar{y} = (1/k)\bar{x} + c. \\
2.\ s_x^2 = k^2 s^2 \quad \text{or} \quad s^2 = (1/k^2) s_x^2.
\end{bmatrix}
$$

The proof of this theorem is omitted, but can be done by combining the procedures used for Theorems 3.1 and 3.2.

Example 3.21
Find the mean and variance for the measurements 100.0, 100.3, 100.8, 100.4, 100.6, and 100.3 using the coding $x_i = 10(y_i - 100)$ for $i = 1, 2, \ldots, 6$.

Solution
The coded measurements are 0, 3, 8, 4, 6, and 3 with

24; 134

$$\Sigma x_i = \underline{\hspace{2cm}} \qquad \Sigma x_i^2 = \underline{\hspace{2cm}}.$$

4

1. $\quad \bar{x} = \dfrac{24}{6} = \underline{\hspace{2cm}}.$

24

7.6

2. $\quad s_x^2 = \dfrac{(134) - \dfrac{(\underline{\hspace{1cm}})^2}{6}}{5} = \underline{\hspace{2cm}}.$

3. Using Theorem 3.3,

100.4

$$\bar{y} = \frac{1}{10}(4) + 100 = \underline{\hspace{2cm}}$$

.076

$$s^2 = \frac{1}{100} s_x^2 = \frac{7.6}{100} = \underline{\hspace{2cm}}.$$

One should remember that coding is to be used only when it is convenient to do so. When a calculator is available, it is often more time consuming to code the data than to do the total calculations in terms of the original measurements. Coding is most efficient when the data consist of fractions in the form of decimals (for example, .0021 and .0035) or large numbers whose last digits are zeros (for example, 16200 and 17500). Use your own judgment in deciding whether to code your data or work with the original measurements.

Self-Correcting Exercises 3E

1. A set of measurements has values recorded to the nearest tenth in which the values all lie between 160.0 and 170.0.
   a. What system of coding would be useful to reduce the arithmetic work involved in calculating the mean and standard deviation for these data?
   b. What would be the relationship between the mean and standard deviation of the original data and the coded data?
2. Calculate $\bar{y}$ and s^2 for the following data using the coding $x_i = y_i - 10$:

$$12, \ 15, \ 14, \ 11, \ 19, \ 16, \ 18, \ 15, \ 14, \ 15.$$

3. The following are the diameters (in inches) of a sample of 10 bolts coming off a production line: .51, .53, .49, .51, .52, .53, .52, .50, .51, .51.
   a. Code the data by first multiplying by 100 and then subtracting 50.
   b. Calculate the mean and variance of the original diameters using the coded measurements.

EXERCISES

1. The following set of data represents the gas mileage for each of 20 cars selected randomly from a production line during the first week in March:

18.1	16.3	18.6	18.7
15.2	19.9	20.3	22.0
19.7	17.7	21.2	18.2
20.9	19.7	19.4	20.2
19.8	17.2	17.9	19.6

   a. What is the range of this data?
   b. Construct a relative frequency histogram for this data using subintervals of width 1.0. (You might begin with 15.15.)
   c. Based on the histogram in b.:
      i. What is the probability that a measurement selected at random from these data will fall in the interval 17.15 to 21.15?
      ii. What is the estimated probability that a measurement taken from the population would be greater than 19.15?
   d. Arrange the measurements in order of magnitude, beginning with 15.2.
   e. What is the median of these data?
   f. The _____th percentile would be any number lying between 16.3 and 17.2.
   g. The _____th percentile would be any number lying between 19.9 and 20.2.
   h. Calculate $\bar{y}$, s^2 and s for these data. (Remember to use the shortcut method.)
   i. Do these data conform to Tchebysheff's Theorem? Support your answer by calculating the fractions of the measurements lying in the intervals $\bar{y} \pm ks$ for $k = 1, 2, 3$.

    j. Does the Empirical Rule adequately describe these data?

2. An experimental strain of "long-stemmed roses" was developed with a mean stem length of 15 inches and standard deviation 2.5 inches.

    a. If one accepts as "long-stemmed roses" only those roses with a stem length greater than 12.5 inches, what percentage of such roses would be unacceptable?

    b. What percentage of these roses would have a stem length between 12.5 and 20 inches?

    Hint: Using the symmetry of the normal distribution, (½ of 68%) of the measurements lie one standard deviation to the left *or* to the right of the mean, and (½ of 95%) of the measurements lie two standard deviations to the left *or* to the right of the mean.

3. The heights of 40 corn stalks ranged from 2.5 ft. to 6.3 ft. In presenting this data in the form of a histogram, suppose you had decided to use 0.5 ft. as the width of your class interval.

    a. How many intervals would you use?

    b. Give the class boundaries for the first and the last classes.

4. A machine designed to dispense cups of instant coffee will dispense on the average μ oz., with standard deviation, $\sigma = .7$ oz. Assume that the amount of coffee dispensed per cup is approximately mound-shaped. If 8 oz. cups are to be used, at what value should μ be set so that approximately 97.5% of the cups filled will not overflow?

5. A pharmaceutical company wishes to know whether an experimental drug being tested in its laboratories has any effect on systolic blood pressure. Fifteen subjects, randomly selected, were given the drug and the systolic blood pressures in millimeters recorded:

172	148	123
140	108	152
123	129	133
130	137	128
115	161	142

    a. Approximate s using the method described in Section 3.9.

    b. Calculate $\bar{y}$ and s for the data.

    c. Find values for the points a and b such that at least 75% of the measurements fall between a and b.

    d. Would Tchebysheff's Theorem be valid if the approximated s were used in place of the calculated s?

    e. Would the Empirical Rule apply to this data?

6. Toss two coins 30 times, recording for each toss the number of heads observed.

    a. Construct a histogram to display the data generated by the experiment.

    b. Find $\bar{y}$ and s for your data.

    c. Do the data conform to Tchebysheff's Theorem? Empirical Rule?

7. The following data represent the social ambivalence scores for 15 people as measured by a psychological test. (The higher the score, the stronger the ambivalence.)

9	8	15	17	10
14	11	4	12	13
10	13	19	11	9

    a. Using the range, approximate the standard deviation, s.

    b. Calculate $\bar{y}$, s^2, and s for this data.

    c. What fraction of the data actually lies in the interval $\bar{y} \pm 2s$?

8. A lumbering company interested in the lumbering rights for a certain tract of slash pine trees is told that the mean diameter of these trees is 14 inches with a standard deviation of 2.8 inches. Assume the distribution of diameters approximately normal.

    a. What fraction of the trees will have diameters between 8.4 inches and 22.4 inches?

    b. What fraction of the trees will have diameters greater than 16.8 inches?

9. If the mean duration of television commercials on a given network is one minute, 15 seconds, with a standard deviation of 25 seconds, what fraction of these commercials would run longer than two minutes, five seconds? Assume that duration times are approximately normally distributed.

Chapter   4

PROBABILITY

4.1 Introduction

We have already stated that our aim is to make inferences about a population based upon sample information. However, in addition to making the inference, we also need to assess how good the inference will be.

Suppose that an experimenter is interested in estimating the unknown mean of a population of observations to within 2 units of its actual value. If an estimate is produced based upon the sample observations, what is the chance that the estimate is no further than 2 units away from the true but unknown value of the mean?

If an investigator has formulated two possible hypotheses about a population and only one of these hypotheses can be true, when the sample data is collected he must decide which hypothesis to accept and which to reject. What is the chance that he will make the correct decision?

In both of these situations, we have used the term "chance" in assessing the goodness of an inference. But chance is just the layman's term for the concept statisticians refer to as _____. Therefore, some elementary results from the theory of probability are necessary to understand how the accuracy of an inference can be assessed.

probability

In the broadest sense, the probability of the occurrence of an event A is a measure of one's belief that the event A will occur in a single repetition of an experiment. One interpretation of this definition that finds widespread acceptance is based upon empirically assessing the probability of the event A by repeating an experiment N times and observing n_A/N, the relative frequency of the occurrence of event A. When N, the number of repetitions, becomes very large, the fraction n_A/N will approach a number we will call $P(A)$, the probability of the occurrence of the event A.

4.2 The Sample Space

A. Sample Points

When the probability of an event must be assessed, it is important that we be able to visualize under what conditions that event will be realized.

experiment
population
sample

An _____ is the process by which an observation or measurement is obtained. When an experiment is run repeatedly, a _____ of observations results. A _____ would consist of any set of observations taken from this population.

A simple event is defined as one of the possible outcomes of a single repetition of the experiment. *One and only one* simple event can occur on a single repetition of an experiment. Simple events are denoted by the letter E with a subscript. Any collection of simple events is called a compound event. Compound events are denoted by capital letters such as A, B, G, and so on.

Example 4. 1
An experiment involves ranking three applicants, X, Y, and Z, in order of their ability to perform in a given position. List the possible simple events associated with this experiment.

Solution
Using the notation (X, Y, Z) to denote the outcome that X is ranked first, Y is ranked second, and Z is ranked third, the six possible outcomes or simple events associated with this experiment are:

X, Y
Z, Y, X

$$E_1: (X, Y, Z) \quad E_2: (Y, X, Z) \quad E_3: (Z, \text{_____})$$
$$E_4: (X, Z, Y) \quad E_5: (Y, Z, X) \quad E_6: (\text{_____})$$

$E_1; E_4$

If A is the event that applicant X is ranked first, then A will occur if simple event _____ or _____ occurs.

Example 4.2
The financial records of two companies are examined to determine whether each company showed a profit (P) or not (N) during the last quarter.
a. List the simple events associated with this experiment.
b. List the simple events comprising the event B: exactly one company showed a profit.

Solution
a. The simple events consist of the *ordered* pairs

N, N

$$E_1: (P, P) \qquad E_2: (P, N)$$
$$E_3: (N, P) \qquad E_4: (\text{_____}).$$

$E_2; E_3$

b. Event B consists of the simple events _____ and _____.

A Venn diagram is a pictorial representation of the possible outcomes of an experiment in which each simple event is associated with a point called a sample point, and all the sample points are enclosed by a closed curve. The totality of the sample points enclosed by the curve is called the _____ _____ and denoted by S.

sample
space

Example 4.3
An oil wildcatter has just enough resources to drill three wells. List the simple

events associated with this experiment and construct a Venn diagram to represent the sample space, S.

Solution
A typical outcome is (gusher, dry, dry) which we will abbreviate to (g, d, d). The simple events are given by the following *ordered triplets:*

E_1: (g, g, g) E_5: (d, d, g)

E_2: (d, g, g) E_6: (d, g, d)

E_3: (g, d, g) E_7: (_____)

E_4: (g, g, d) E_8: (_____).

g, d, d

d, d, d

Complete the following Venn diagram corresponding to this experiment by assigning the eight simple events to the eight points enclosed by the closed curve:

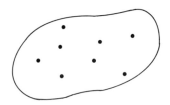   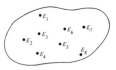

B. The Assignment of Probabilities
When the sample space has been defined, the next step is to assign probabilities to each of the simple events. Using the relative frequency interpretation of the probability of an event imposes two restrictions on the assignment of probabilities to the simple events. These are

a.   _____ $\leq P(E_i) \leq$ _____ .   0; 1

b. $\sum_{\text{all } i} P(E_i) =$ _____ .   1

After the probabilities have been assigned to the simple events $E_1, E_2, \ldots,$ then the probability associated with any compound event A is found by summing the probabilities of all the simple events that comprise the event A:

$$P(A) = \sum_{\text{all } E_i \text{ in } A} P(E_i).$$

If an experiment has N possible equally likely outcomes, then the condition

$$\sum_{i=1}^{N} P(E_i) = 1$$

1/N

requires that $P(E_i) =$ _____ for $i = 1, 2, \ldots, N$. In this case, if the event A contains n_A sample points, then

$$P(A) = \frac{n_A}{N}.$$

Remember that this is a special case and the probabilities assigned to the sample points will not in general be equal.

Example 4.4

Suppose that two coins are tossed and the upper faces recorded. Suppose further that the coins are not fair and the probability that a head results on either coin is greater than a half. The following probabilities are assigned to the sample points:

Sample Point	Outcome	Probability
E_1	HH	.42
E_2	HT	.18
E_3	TH	.28
E_4	TT	.12

a. Verify that this assignment of probabilities satisfies the conditions

$$1.\ 0 \leqslant P(E_i) \leqslant 1.$$

$$2.\ \sum_{i=1}^{4} P(E_i) = 1.$$

b. Find the probability of the event A: the toss results in exactly one head and one tail.
c. Find the probability of the event B: the toss results in at least one head.

Solution

a. We need but verify (2) since observation shows that the assigned probabilities satisfy condition (1). Hence,

$$\sum_{i=1}^{4} P(E_i) = P(E_1) + P(E_2) + P(E_3) + P(E_4)$$

.28; .12

$$= .42 + .18 + \underline{\hspace{1cm}} + \underline{\hspace{1cm}}$$

1

$$= \underline{\hspace{1cm}}$$

is

and condition (2) (is, is not) satisfied.

b. The event A: exactly one head and one tail, consists of the sample points E_2 and E_3. Therefore,

$$P(A) = P(E_2) + P(E_3)$$

$$= \underline{\hspace{1.5cm}} + \underline{\hspace{1.5cm}}$$

.18; .28

$$= \underline{\hspace{1.5cm}}.$$

.46

c. The event B: at least one head, consists of the sample points E_1, E_2, and E_3. Therefore,

$$P(B) = P(E_1) + P(E_2) + P(E_3)$$

$$= \underline{\hspace{1.5cm}} + .18 + .28$$

.42

$$= \underline{\hspace{1.5cm}}.$$

.88

Example 4.5
In a shipment of four radios, R_1, R_2, R_3, and R_4, one radio is defective (say R_3). If a dealer selects two radios at random to display in his store, what is the probability that exactly one of the radios is defective?

Solution
If we disregard the order of selection of the two radios to be displayed, the possible outcomes are:

$$E_1: (R_1, R_2) \qquad E_4: (R_2, R_3)$$
$$E_2: (R_1, R_3) \qquad E_5: (R_2, \underline{\hspace{1cm}})$$
$$E_3: (R_1, R_4) \qquad E_6: (\underline{\hspace{1cm}}).$$

R_4

R_3, R_4

If the radios are selected at random, all combinations should have the same chance of being drawn. Therefore, we assign $P(E_i) = \underline{\hspace{1.5cm}}$ to each of the six sample points.

1/6

The event D: exactly one of the two radios selected is defective, consists of the sample points E_2, E_4, and $\underline{\hspace{1.5cm}}$. Therefore,

E_6

$$P(D) = P(E_2) + P(E_4) + P(E_6)$$

$$= 1/6 + 1/6 + 1/6$$

$$= \underline{\hspace{1.5cm}}.$$

1/2

Self-Correcting Exercises 4A

1. A lot containing six items is comprised of four good items and two defective items. Two items are selected at random from the lot for testing purposes.
   a. List the sample points for this experiment.
   b. List the sample points in each of the three following events:
      A: at least one item is defective.
      B: exactly one item is defective.

C: no more than one item is defective.

2. A hospital spokesman reported that four births had taken place at the hospital during the last twenty-four hours. If we consider only the sex of these four children, recording M for a male child and F for a female child, there are sixteen sex combinations possible.

   a. List these sixteen outcomes in terms of sample points, beginning with E_1 as the outcome (FFFF).

   b. Define the following events in terms of the sample points $E_1, \ldots, E_{16}$:

         A: two boys and two girls are born.

         B: no boys are born.

         C: at least one boy is born.

         D: either A or B occurs.

         E: both A and C occur.

         F: either A or C or both A and C occur.

   c. If the sex of a newborn baby is just as likely to be male as female, find the probabilities associated with the six events defined in b.

4.3 Results Useful in Counting Sample Points

There are three basic counting rules that are useful in counting the number of sample points, N, arising in many experiments. When all the N sample points are equally likely, the probability of an event A can be found without listing the sample points if N, the number of points in S, and n_A, the number of points in A, can be counted, since in this case, $P(A) = n_A/N$.

> *The mn rule.* If a procedure can be completed in two stages and the first stage can be done in m ways and the second stage in n ways after the first stage has been completed, then the number of ways of completing the procedure is mn (m times n).

Example 4.6

An experiment involves ranking three applicants in order of merit. How many ways can the three applicants be ranked?

Solution

The process of ranking three applicants can be accomplished in two stages.

    Stage 1: Select the best applicant from the three.

    Stage 2: Having selected the best, select the next best from the remaining two applicants.

The ranking of the remaining applicant will automatically be third. The number of ways of accomplishing stage 1 is _____. When stage 1 is completed, there are _____ ways of accomplishing stage 2. Hence there are (3)(2) = _____ ways of ranking three applicants.

3
2
6

Example 4.7

A lot of items consists of four good items (G_1, G_2, G_3, and G_4) and two defective items (D_1 and D_2).

a. How many different samples of size two can be formed by selecting two items from these six?

b. How many different samples will consist of exactly one good and one defective item?

c. What is the probability that exactly one good and one defective will be drawn?

Solution

a. Selecting two items from six items corresponds to the two-step procedure of (1) picking the first item and (2) picking the second item after picking the first. Hence $m =$ _____, $n =$ _____, and the number of ordered pairs is $N = mn = (6)(5) =$ _____.

6; 5
30

b. Selecting one good and one defective item can be done in either of two ways.

  1. The *defective* item can be drawn *first* in $m =$ _____ ways and the *good* item drawn *second* in $n =$ _____ ways. Hence there are $mn =$ _____ ways of selecting a defective item on the first draw and a good item on the second draw.

2
4
8

  2. However, the *good* item can be drawn *first* in $m =$ _____ ways and the *defective* item drawn second in $n =$ _____ ways, so that there are $mn =$ _____ ways in which a good item is drawn first and a defective item is drawn second.

4
2
8

  3. Combining the results of (1) and (2), there are exactly $8 + 8 =$ _____ samples that will contain exactly one defective and one good item.

16

c. Let A be the event that exactly one good and one defective item are drawn. From part a., $N =$ _____, and from part b., $n_A =$ _____.
Hence

30; 16

$$P(A) = \frac{n_A}{N} = \frac{16}{30} = \underline{\hspace{2cm}}.$$

8/15

Permutations. An ordered arrangement of r distinct objects is called a permutation. The number of permutations consisting of r objects selected from n objects is given by the formula

$$P_r^n = n(n-1)(n-2)\ldots(n-r+1).$$

Notice that P_r^n consists of r factors commencing with n.

Example 4.8

In how many ways can three different office positions be filled if there are seven applicants who are qualified for all three positions?

Solution

Notice that assigning the same three people to different office positions would produce different ways of filling the three positions. Hence we need to find the number of permutations (*ordered arrangements*) of three people selected from seven. Therefore,

$$P_3^7 = (7)(6)(\underline{\hspace{1.5cm}}) = \underline{\hspace{1.5cm}}.$$

5; 210

Example 4.9

A corporation will select two sites from ten available sites under consideration for building two manufacturing plants. If one plant will produce flashbulbs and the other cameras, in how many ways can the selection be made?

Solution
We are interested in the number of permutations of two sites selected from ten sites, since if two sites, say 6 and 8, were chosen, and the flashbulb plant was built at site 6 while the camera plant was built at site 8, this would result in a different selection than would occur if the camera plant was built at site 6 and the flashbulb plant at site 8. Therefore, the number of selections is

$$P_2^{10} = (10) \,(\underline{\hspace{2cm}}) = \underline{\hspace{2cm}}.$$

Combinations. A selection of r objects from n distinct objects without regard to their ordering is called a combination. The number of combinations that can be formed when selecting r objects from n objects is given as

$$\left[\quad C_r^n = \frac{n!}{r! \,(n-r)!} \quad \right].$$

Example 4.10
How many different five-card hands can be dealt from an ordinary deck of 52 cards?

Solution
Since it is the value of the five cards and not the order in which they were dealt that will differentiate one five-card hand from another, the number of distinct five-card hands is

2,598,960

$$C_5^{52} = \frac{52!}{5! \,47!} = \frac{(52)\,(51)\,(50)\,(49)\,(48)}{(5)\,(4)\,(3)\,(2)\,(1)} = \underline{\hspace{2cm}}.$$

Notice that C_5^{52} is the same as C_{47}^{52}. In general

$$C_r^n = \underline{\hspace{2cm}}.$$

Example 4.11
An experimenter must select three animals from ten available animals to be used as a control group. In how many ways can the control group be selected?

Solution
Since the order of selection is unimportant, the number of unordered selections is

120

$$C_3^{10} = \frac{(10)\,(9)\,(8)}{(3)\,(2)\,(1)} = \underline{\hspace{2cm}}.$$

4.4   Calculating the Probability of an Event: Sample Point Approach

The sample point approach to finding $P(A)$ comprises five steps, which are:

Step 1. Define the experiment.
Step 2. List all the sample points. Test to make certain that none can be decomposed.
Step 3. Specify which sample points lie in A.
Step 4. Assign appropriate probabilities to the sample points. Make sure that

$$\sum_{\text{all } i} P(E_i) = 1.$$

Step 5. Find $P(A)$ by summing the probabilities for all points in A.

Example 4.12

A lot of five items includes three good items, G_1, G_2, and G_3, and two defective items, D_1 and D_2. Two items are selected at random from this lot. We shall find the probability of the event A, that at least one of the two items selected is defective.

Solution

Step 1. The experiment is the selection at random of a pair of items from among _____ distinct items.   five

Step 2. The sample space contains the $C_2^5 =$ _____ sample points   10
$(G_1G_2), (G_1G_3), (G_1D_1), (G_1D_2), (G_2G_3), (G_2D_1), (G_2D_2),$
$(G_3D_1), (G_3D_2),$ and $(D_1D_2).$

Step 3. $A = \{(G_1D_1), (G_1D_2), (G_2D_1), (G_2D_2), (G_3D_1), (\underline{\quad\quad}),$ G_3D_2
$(\underline{\quad\quad})\}.$ D_1D_2

Step 4. The requirement that the sampling be random implies that the same probability _____ should be assigned to each sample point.   1/10

Step 5. There are _____ sample points in A. Hence $P(A) =$ _____.   seven; 7/10

Modified Sample Point Approach. It may be that the list of sample points is quite long. But if equal probabilities are assigned to the sample points, all that is actually required is that the student know precisely the number, N, of points in S and the number, n_A, of points in the event A. Then $P(A) = n_A/N$. If, however, a list is not made of the sample points, the student must take care that no sample point in A is overlooked.

Example 4.13

A dealer who buys items in lots of ten selects three of the ten items at random and inspects them thoroughly. He accepts all ten if there are no defectives among the three inspected. Suppose that a lot contains two defective items.
a. What is the probability that the dealer will nonetheless accept all ten?
b. What is the probability that he will find both of the defectives?

Solution

a. Let A be the event that the dealer accepts the lot.
   Step 1. The experiment consists of selecting three items at random from ten items.
   Step 2. The sample space consists of *unordered* triples of the form $(G_1G_2G_8)$ or $(G_7D_1D_2)$. The number of sample points is

$$N = C_3^{10} = \frac{(10)(9)(8)}{(3)(2)(1)} = \underline{\quad\quad}.$$ 120

56

1/120

56/120

8

8/120

1

8
8

Step 3. The event A consists of the sample points containing three good items. Hence

$$n_A = C_3^8 = \frac{(8)\,(7)\,(6)}{(3)\,(2)\,(1)} = \underline{\qquad}.$$

Step 4. Since the selection is at random, each sample point should be assigned the same probability equal to _____ .

Step 5. There are 56 sample points in A. Hence $P(A) = \underline{\qquad}$.

b. Let B be the event that the dealer finds both defective items in the random selection.

Step 3. The event B consists of the sample points $(D_1 D_2 G_1)$, $(D_1 D_2 G_2)$, $(D_1 D_2 G_3)$, ..., $(D_1 D_2 G_8)$. Hence there are $n_B = \underline{\qquad}$ sample points in B.

Step 5. Therefore, $P(B) = \dfrac{n_B}{N} = \underline{\qquad}$.

Notice that we can count n_B in the following way. The selection of the three items can be thought of as occurring in two stages:

Stage 1: The two defectives are selected. There are $m = C_2^2 = \underline{\qquad}$ ways to do this.

Stage 2: The remaining nondefective is selected. There are $n = C_1^8$ = _____ ways to do this.

Hence there are $mn = (1)\,(8) = \underline{\qquad}$ ways to select two defectives and one nondefective from the ten items consisting of two defectives and eight nondefectives.

Self-Correcting Exercises 4B

1. The owner of a camera shop receives a shipment of five cameras from a camera manufacturer. Unknown to the owner, two of these cameras are defective. Suppose that the owner selects two of the five cameras at random and tests them for operability.
   a. Define the experiment.
   b. List the sample points associated with this experiment.
   c. Define the following events in terms of the sample points in b.
      A: Both cameras are defective.
      B: Neither camera is defective.
      C: At least one camera is defective.
      D: The first camera tested is defective.
   d. Find $P(A), P(B), P(C)$, and $P(D)$ using the sample point approach.
2. Refer to Exercise 1.
   a. Use counting rules to count the number of sample points in the experiment.
   b. Count the number of sample points in the events defined in part c. of Exercise 1.

c. Find $P(A), P(B), P(C),$ and $P(D)$ using the modified sample point approach. Compare your answers with part d. of Exercise 1.

3. In quality control of taste and texture, it is common to have a taster compare a new batch of a food product with one having the desired properties. Three new batches are independently tested against the standard and classified as having the desired properties (H) or not having the desired properties (N).

   a. List the sample points for this experiment. Use a counting rule to verify that all sample points have been listed.

   b. If in fact all three new batches are no different from the standard, all sample points in a. should be equally likely. If this is the case, find the probabilities associated with the following events:

   > A: Exactly one batch is declared as not having the desired properties.
   > B: Batch number one is declared to have the desired properties.
   > C: All three batches are declared to have the desired properties.
   > D: At least two batches are declared to have the desired properties.

4.5 Compound Events

When attempting to find the probability of an event A, it is often useful and convenient to express A in terms of other events whose probabilities are known or perhaps easily calculated. Composition of events occurs in one of the two following ways or a combination of these two:

> *Intersections.* The intersection of two events A and B is the event consisting of those sample points which are in both A and B. The intersection of A and B is denoted by AB.

> *Unions.* The union of two events A and B is the event consisting of those sample points that are in either A or B or both A and B. The union of A and B is denoted by $A \cup B$.

Example 4.14
In each of the Venn diagrams following, express symbolically the event represented by the shaded area. In each case the sample space S comprises all sample points within the rectangle.

a.

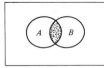

AB

b.

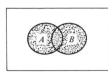

Symbol_____

$A \cup B$

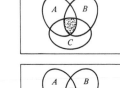

Symbol_____

c.

ABC

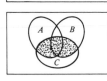

Symbol_____

d.

$AC \cup BC$

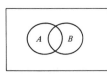

Symbol_____

Example 4.15

In each of the Venn diagrams below shade in the event symbolized.

a. Symbol: $A \cup B$

b. Symbol: BC

c. Symbol: $AE_1 \cup AE_2$

A

d. Note that $E_1 \cup E_2 = S$ and $AE_1 \cup AE_2 =$ _____.

4.6 Event Relations

Events may be related to other events in several ways. Relations between events can often be used to simplify calculations involved when finding the probability of an event.

> The event consisting of all those sample points in the sample space S that are not in the event A is defined as the *complement of A* and is denoted by $\bar{A}$.

It is always true that $P(A) + P(\bar{A}) =$ _____ . Therefore, $P(A) = 1 - P(\bar{A})$.
If $P(\bar{A})$ can be found more easily than $P(A)$, this relationship greatly simplifies finding $P(A)$.

Example 4.16
If three fair coins are tossed, what is the probability of observing at least one head in the toss?

Solution
a. Let A be the event that there is at least one head in the toss of three coins. $\bar{A}$ is the event that there are _____ heads in the toss of three coins.
b. There are $N = 8$ possible outcomes for this experiment so that each sample point would be assigned a probability of _____ .
c. $\bar{A}$ consists of the single sample point (TTT); $P(\bar{A}) =$ _____ and $P(A)$ $= 1 - P(\bar{A}) =$ _____ .

1

no

1/8
1/8
7/8

Events A and B which have no common sample points are said to be *mutually exclusive*.

When A and B are mutually exclusive, their intersection AB contains no sample points and $P(AB) =$ _____ . Notice that the events A and $\bar{A}$ (are, are not) mutually exclusive.

0

are

To introduce the concept of independence of two events, it is first necessary to define a conditional probability.

The probability that event A has occurred, *given that* the event B has occurred, is given by

$$P(A|B) = \frac{P(AB)}{P(B)}$$

for $P(B) > 0$.

Use a Venn diagram with events A and B to see that by knowing the event B has occurred, you effectively exclude any sample points lying outside of the event B from further consideration. Since $P(B)$ and $P(AB)$ represent the amounts of probability associated with events B and AB, then

$$P(A|B) = \frac{P(AB)}{P(B)}$$

merely represents the proportion of $P(B)$ that will give rise to the event A.
When $P(A|B) = P(A)$ the events A and B are said to be (probabilistically) independent, since the probability of the occurrence of A is not affected by knowledge of the occurrence of B. If $P(A|B) \neq P(A)$, the events A and B are said to be dependent.

Example 4.17
You hold ticket numbers 7 and 8 in an office lottery in which ten tickets

numbered 1 through 10 were sold. The winning ticket is drawn at random from those sold. You are told that the winning number is odd. Does this information alter the probability that you have won the lottery or are the two events independent?

Solution

Define the events A and B as follows:

A: Number 7 or 8 is drawn.

B: An odd number is drawn.

1/5

1/5

independent

The unconditional probability of winning is $P(A) = 2/10 =$ _____,
while the conditional probability of winning is $P(A|B) =$ _____. Your probability of winning remains unchanged; the events A and B are (dependent, independent).

Self-Correcting Exercises 4C

1. Refer to Exercise 1, Self-Correcting Exercises 4B.
   a. List the sample points comprising the following events: $A \cup C$, AD, CD and $A \cup D$.
   b. Calculate $P(A \cup C)$, $P(AD)$, $P(CD)$, $P(A \cup D)$.
2. Refer to Exercise 2, Self-Correcting Exercises 4A.
   a. Rewrite events D, E, and F in terms of the events A, B, and C.
   b. List the sample points in the following events: AB, $B \cup C$, $AC \cup BC$, $\bar{C}$, $\overline{AC}$.
   c. Using the results of part c. of Exercise 2, Self-Correcting Exercises 4A, calculate $P(A \cup B)$, $P(\bar{C})$ and $P(\overline{AC})$.
   d. Calculate $P(A|C)$. Are A and C mutually exclusive? Are A and C independent?
   e. Calculate $P(B|C)$. Are B and C independent? Mutually exclusive?
3. Two hundred corporate executives in the Los Angeles area were interviewed. They were classified according to the size of the corporation they represented and their choice as to the most effective method for reducing air pollution in the Los Angeles basin. (Data are fictitious.)

		Corporation Size	
Option	Small (A)	Medium (B)	Large (C)
Car Pooling (D)	20	15	20
Bus Expansion (E)	30	25	11
Gas Rationing (F)	3	8	4
Conversion to Natural Gas (G)	10	7	5
Anti-pollution Devices (H)	12	20	10

Suppose one executive is chosen at random to be interviewed on a television broadcast.

   a. Calculate the following probabilities and describe each event involved in terms of the problem: $P(A)$, $P(F)$, $P(AF)$, $P(A \cup G)$, $P(AD)$ and $P(\bar{F})$.
   b. Calculate $P(A|F)$ and $P(A|D)$. Are A and F independent? Mutually exclusive? Are A and D independent? Mutually exclusive?

4.7 Two Probability Laws and Their Use

The additive law of probability for two events A and B is given by

$$P(A \cup B) = P(A) + P(B) - P(AB).$$

When the events A and B are mutually exclusive, $P(AB) = $ _____ and   0
the additive law becomes

$$P(A \cup B) = P(A) + P(B).$$

If the event A is contained in the event B, then $P(A \cup B) = $ _____. $P(B)$
   The multiplicative law of probability for two events A and B is given by

$$P(AB) = P(A)P(B|A)$$

or

$$P(AB) = P(B)P(A|B).$$

If the events A and B are independent, then

$$P(AB) = P(A)P(B).$$

Example 4.18
In the following Venn diagram the ten sample points shown are equally likely.
Thus, to each sample point is assigned the probability _____.   1/10

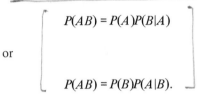

$P(A) = \frac{4}{10} = 2/5$

a. $P(A \cup B) = $ _____.   7/10
b. $P(AB) = $ _____.   2/10
c. $P(B) = $ _____.   5/10

$\dfrac{P(AB)}{P(B)} = \dfrac{\frac{2}{10}}{\frac{5}{10}} = \frac{2}{5} = P(A)$

d. $P(A|B) = $ _____.   2/5
e. $P(\bar{B}) = $ _____.   5/10
f. A and B are (independent, dependent).   independent

Example 4.19
The personnel files for the Quick Sales Agency lists its 150 employees as
follows:

	Years Employed with Quick		
	0–5 (A)	6–10 (B)	11 or more (C)
Not a College Graduate (D)	10	20	20
College Graduate (E)	40	50	10

If *one* personnel file is drawn at random from Quick's personnel files, calculate the probabilities requested below:

1/3   a. $P(A) =$ _____.

2/3   b. $P(E) =$ _____.

2/15   c. $P(BD) =$ _____.

1/10   d. $P(C/E) =$ _____.

11/15   e. $P(A \cup E) =$ _____.

0   f. $P(A/C) =$ _____.

dependent   g. A and E are (independent, dependent).

4.8 Calculating the Probability of an Event: Event Composition Approach

The event composition approach to finding $P(A)$ comprises four steps, which are:

Step 1. Define the experiment.

Step 2. Clearly visualize the nature of the sample points. Identify a few to clarify your thinking.

Step 3. Write an equation expressing A as a composition of two or more events. Make certain that the event expressed by the composition is the same set of points as the event A.

Step 4. Apply the additive and multiplicative laws of probability as required to the equation found in step 3. It is assumed that the component probabilities are known for the particular composition used.

Example 4.20

Player A has entered a tennis tournament but it is not yet certain whether Player B will enter. Let A be the event that Player A will win the tournament and let B be the event that Player B will enter the tournament. Suppose that Player A has probability 1/6 of winning the tournament if Player B enters and probability 3/4 of winning if Player B does not enter the tournament. If $P(B) = 1/3$, find $P(A)$.

Solution

Step 1.   The experiment is the observation of whether Player B enters the tournament and whether Player A wins the tournament.

Step 2.   The sample space S can be partitioned into four parts, AB, $A\bar{B}$, $\bar{A}B$, and $\bar{A}\bar{B}$. Recall that $\bar{B}$ is the event that B does not occur, called the

complement   _____ of B, and hence $A\bar{B}$ would be the event that A occurs and B does not occur.

Step 3.   Now $B \cup \bar{B} = S$ and $A = A \cap (B \cup \bar{B})$. Therefore $A = AB \cup A\bar{B}$. (Use a Venn diagram to verify this result if you wish.)

Step 4.   The events AB and $A\bar{B}$ are mutually exclusive since both events cannot occur simultaneously. Therefore,

$$P(A) = P(AB \cup A\bar{B})$$

$$= P(AB) + P(A\bar{B}).$$

But

$$P(AB) = P(B)P(A|B)$$

$$= (1/3) (\underline{\hspace{2cm}})$$ 1/6

$$= \underline{\hspace{2cm}}$$ 1/18

and

$$P(A\bar{B}) = P(\bar{B})P(A|\bar{B})$$

$$= (2/3) (\underline{\hspace{2cm}})$$ 3/4

$$= \underline{\hspace{2cm}}.$$ 1/2

Therefore, $P(A) = \underline{\hspace{2cm}} + \underline{\hspace{2cm}} = \underline{\hspace{2cm}}.$ 1/18; 1/2; 5/9

Example 4.21

The selling style of a temperamental salesman is strongly affected by his success or failure in the preceding attempt to sell. If he has just made a sale, his confidence and effectiveness rise and the probability of selling to his next prospect is 3/4. When he fails to sell, his manner is fearful and the probability that he sells to his next prospect is only 1/3. Suppose that the probability he will sell to his first contact on a given day is 1/2. We shall find the probability of the event A, that he makes at least two sales on his first three contacts on a given day.

Solution

Step 1. The experiment is the observation of this salesman's successes and failures on his first three contacts of the day.

Step 2. As in the preceding example, each sample point is the intersection of compound events. Let S_1 and N_1 denote respectively the events "sale on the first contact" and "no sale on the first contact." Similarly, S_2, N_2, S_3, and N_3 shall denote "sale" and "no sale" on the second and third contacts. Then the eight sample points are $S_1 S_2 S_3$, $S_1 S_2 N_3$, $S_1 N_2 S_3$, $N_1 S_2 S_3$, $S_1 N_2 N_3$, $N_1 S_2 N_3$, $N_1 N_2 S_3$, and $N_1 N_2 N_3$.

Step 3. Since A is the event that at least two sales were made on the first three contacts,

$$A = S_1 S_2 S_3 \cup S_1 S_2 N_3 \cup S_1 N_2 S_3 \cup \underline{\hspace{2cm}}.$$ $N_1 S_2 S_3$

Step 4. Each of the sample points comprising A is the intersection of three events which are not independent. Notice that for three events B, C, and D, that

$$P(BCD) = P(BC)P(D|BC)$$

$$= P(B)P(C|B)P(D|BC).$$

Therefore,

$$P(S_1 S_2 S_3) = P(S_1)P(S_2|S_1)P(S_3|S_1 S_2)$$

9/32

$$= (1/2)\,(3/4)\,(3/4) = \underline{\qquad}$$

3/32

$$P(S_1 S_2 N_3) = (1/2)\,(3/4)\,(1/4) = \underline{\qquad}$$

1/3; 1/24

$$P(S_1 N_2 S_3) = (1/2)\,(1/4)\,(\underline{\qquad}) = \underline{\qquad}$$

1/3; 3/4; 3/24

$$P(N_1 S_2 S_3) = (1/2)\,(\underline{\qquad})\,(\underline{\qquad}) = \underline{\qquad}$$

and

13/24

$$P(A) = 9/32 + 3/32 + 1/24 + 3/24 = \underline{\qquad}.$$

Self-Correcting Exercises 4D

1. An investor holds shares in three independent companies which, according to his business analyst, should show an increase in profit per share with probabilities 0.4, 0.6, and 0.7 respectively. Assuming that the analyst's estimates for the probabilities of profit increases are correct,
   a. Find the probability that all three companies show profit increases for the coming year.
   b. Find the probability that none of the companies shows a profit increase.
   c. Find the probability that at least one company shows a profit increase.
2. A student prepares for an exam by studying a list of ten problems. He can solve six of the ten problems. For the exam, his professor selects five problems at random from the list of ten problems.
   a. What is the probability that the student can solve all five problems on the exam?
   b. What is the probability that the student passes the exam if he must correctly answer at least three questions to get a passing grade?
   c. What is the probability that the student fails the exam?

4.9 Bayes' Rule (Optional)

A very interesting application of conditional probability is found in Bayes' rule. This rule assumes that the sample space can be partitioned into k mutually exclusive events (subpopulations) $H_1, H_2, \ldots, H_k$ so that $S = H_1 \cup H_2 \cup H_3 \cup \ldots \cup H_k$. If a single repetition of an experiment results in the event A with $P(A) > 0$, we may be interested in making an inference as to which event (subpopulation) most probably gave rise to the event A. The probability that the subpopulation H_i was sampled, given that event A occurred is

$$P(H_i|A) = \frac{P(H_i A)}{P(A)}$$

$$= \frac{P(H_i)P(A|H_i)}{\displaystyle\sum_{j=1}^{k} P(H_j)P(A|H_j)} .$$

The following Venn diagram with $k = 3$ subpopulations demonstrates that

$$P(A) = P(AH_1 \cup AH_2 \cup AH_3)$$

$$= P(AH_1) + P(AH_2) + P(AH_3).$$

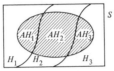

Since $P(AH_i) = P(H_i)P(A|H_i)$, we can write

$$P(A) = P(H_1)P(A|H_1) + P(H_2)P(A|H_2) + P(H_3)P(A|H_3).$$

Hence for k subpopulations,

$$P(A) = \sum_{j=1}^{k} P(H_j)P(A|H_j).$$

In order to apply Bayes' rule, it is necessary to know the probabilities $P(H_1)$, $P(H_2), \ldots , P(H_k)$. These probabilities are called the _____   prior
probabilities. When the prior probabilities are unknown, it is possible to
assume that all subpopulations are equally likely so that $P(H_1) = \ldots = P(H_k)$
$= 1/k$. The conditional probabilities $P(H_i|A)$ are called the _____   posterior
probabilities since these are the probabilities which result after taking account
of the sample information contained in the event A.

Example 4.22
A manufacturer of air-conditioning units purchases 70% of his thermostats
from Company A, 20% from Company B, and the rest from Company C.
Past experience shows that 0.5% of Company A's thermostats, 1% of Company B's thermostats, and 1.5% of Company C's thermostats are likely to be
defective. An air-conditioning unit randomly selected from this manufacturer's
production line was found to have a defective thermostat.
a. Find the probability that the defective thermostat was supplied by
   Company A.
b. Find the probability that the defective thermostat was supplied by
   Company B.

Solution
Let A, B, and C be the events that a thermostat selected at random was supplied by Company A, B, or C respectively. Let D be the event that a defective

thermostat is observed. (Notice that $A \cup B \cup C = S$ and A, B, and C are mutually exclusive.) The following information is available.

$$P(A) = .7 \qquad\qquad P(D|A) = .005$$

.2 $\qquad$ $$P(B) = \underline{\hspace{2cm}} \qquad\qquad P(D|B) = .010$$

.1 $\qquad$ $$P(C) = \underline{\hspace{2cm}} \qquad\qquad P(D|C) = .015.$$

a. Using Bayes' rule to find $P(A|D)$, we have

$$P(A|D) = \frac{P(A)P(D|A)}{P(A)P(D|A) + P(B)P(D|B) + P(C)P(D|C)}$$

$$= \frac{(.7)(.005)}{(.7)(.005) + (.2)(.010) + (.1)(.015)}$$

$$= \frac{.0035}{.0035 + .0020 + .0015}$$

1/2 $\qquad$ $$= \underline{\hspace{2cm}}.$$

b. Using the results of part a. to find $P(B|D)$ we have

$$P(B|D) = \frac{P(B)P(D|B)}{P(A)P(D|A) + P(B)P(D|B) + P(C)P(D|C)}$$

.0020 $\qquad$ $$= \frac{\overline{\hspace{1.5cm}}}{.0070}$$

2/7 $\qquad$ $$= \underline{\hspace{2cm}}.$$

3/14 $\qquad$ Further, $P(C|D) = 1 - 1/2 - 2/7 = \underline{\hspace{2cm}}$. If one had to make a decision as to which company most probably supplied the defective part,

A $\qquad$ Company $\underline{\hspace{2cm}}$ would be so named.

It is interesting to notice that Bayes' rule entails deductive rather than inductive reasoning. Usually we are interested in investigating a problem beginning with the cause and reasoning to its effect. Bayes' rule, however, reasons from the effect (a defective is observed) to the cause (which population produced the defective?). The next example further illustrates this type of logic.

Example 4.23
Suppose that a transmission system, whose input X is either a zero or a one and whose output Y is either a zero or a one, mixes up the input according to the following scheme:

$$P(Y = 0 \mid X = 0) = .90$$

$$P(Y = 1 \mid X = 0) = .10$$

$$P(Y = 0 \mid X = 1) = .15$$

$$P(Y = 1 \mid X = 1) = .85.$$

If $P(X = 0) = .3$, find
a. $P(X = 1 \mid Y = 1)$.
b. $P(X = 0 \mid Y = 0)$.

Solution
Since $P(X = 0) = .3$, $P(X = 1) = 1 - .3 =$ _____.   .7
a. To find $P(X = 1 \mid Y = 1)$, write

$$P(X = 1 \mid Y = 1) = \frac{P(X = 1, Y = 1)}{P(Y = 1)}.$$

The numerator is

$$P(X = 1, Y = 1) = P(X = 1)P(Y = 1 \mid X = 1)$$

$$= (.7)(.85)$$

$$= \underline{\hspace{2cm}}.$$ .595

The denominator is

$$P(Y = 1) = P(X = 0, Y = 1) + P(X = 1, Y = 1)$$

$$= P(X = 0)P(Y = 1 \mid X = 0) + P(X = 1)P(Y = 1 \mid X = 1)$$

$$= (.3)(.10) + (.7)(.85)$$

$$= \underline{\hspace{2cm}}.$$ .625

Therefore,

$$P(X = 1 \mid Y = 1) = \frac{.595}{.625} = \underline{\hspace{2cm}}.$$ .952

b. To find $P(X = 0 \mid Y = 0)$, write

$$P(X = 0 \mid Y = 0) = \frac{P(X = 0, Y = 0)}{P(Y = 0)}.$$

The numerator is

$$P(X = 0, Y = 0) = P(X = 0)P(Y = 0 \mid X = 0)$$

.90

$$= (.3) (\underline{\hspace{2cm}})$$

.270

$$= \underline{\hspace{2cm}}.$$

The denominator is

$$P(Y = 0) = P(X = 0, Y = 0) + P(X = 1, Y = 0)$$

$$= P(X = 0)P(Y = 0 \mid X = 0) + P(X = 1)P(Y = 0 \mid X = 1)$$

.7; .15

$$= (.3) (.90) + (\underline{\hspace{2cm}}) (\underline{\hspace{2cm}})$$

.375

$$= \underline{\hspace{2cm}}.$$

Therefore,

.720

$$P(X = 0 \mid Y = 0) = (.270)/(.375) = \underline{\hspace{2cm}}.$$

Self-Correcting Exercises 4E

1. An oil wildcatter must decide whether or not to hire a seismic survey before deciding whether to drill for oil on a plot of land. Given that oil is present, the survey will indicate a favorable result with probability 0.8; if oil is not present, a favorable result will occur with probability 0.3. The wildcatter figures that oil is present on the plot of land with probability equal to 0.5. Determine the effectiveness of the survey by computing the probability that oil is present on the land, given a favorable seismic survey outcome and comparing this posterior probability with the prior probability of finding oil on the plot of land.

2. Each item coming off a given production line is inspected by either Inspector 1 or Inspector 2. Inspector 1 inspects about 60% of the production items, while Inspector 2 inspects the rest. Inspector 1, who has been at his present job for some time, will not find 1% of the defective items he inspects. Inspector 2, who is newer on the job, misses about 5% of the defective items he inspects. If an item that has passed an inspector is found to be defective, what is the probability that it was inspected by Inspector 1?

EXERCISES

1. Suppose that an experiment requires the ranking of three applicants, A, B, and C, in order of their abilities to do a certain job. The sample points could then be symbolized by the ordered triplets, (ABC), (BAC), and so on.

a. The event A that applicant A will be ranked first comprises which of the sample points?

b. The event B that applicant B will be ranked third comprises which of the sample points?

c. List the points in $A \cup B$.

d. List the points in AB.

e. If equal probabilities are assigned to the sample points show whether or not events A and B are independent.

2. An antique dealer had accumulated a number of small items including a valuable stamp collection and a solid gold vase. To make room for new stock he distributed these small items among four boxes. Without revealing which items were placed in which box, the dealer stated that the stamp collection was included in one box and the gold vase in another. The four boxes were sealed and placed on sale, each at the same price. A certain customer purchased two boxes selected at random from the four boxes. What is the probability that he acquired:

a. the stamp collection?

b. the vase?

c. at least one of these bonus items?

3. If the probability that an egg laid by an insect hatches is $p = .4$, what is the probability that at least three out of four eggs will hatch?

4. In the past history of a certain serious disease it has been found that about 1/2 of its victims recover.

a. Find the probability that exactly one of the next five patients suffering from this disease will recover.

b. Find the probability that at least one of the next five patients suffering from this disease will recover.

5. The sample space for a given experiment comprises the simple events, E_1, E_2, E_3, and E_4. Let the compound events, A, B, and C, be defined by equations:

$$A = E_1 \cup E_2, \quad B = E_1 \cup E_4, \quad C = E_2 \cup E_3.$$

Construct a Venn diagram showing the events, E_1, E_2, E_3, E_4, A, B, and C.

6. Refer to Exercise 5. Probabilities are assigned to the simple events as indicated in the following table:

Simple Event	E_1	E_2	E_3	E_4
Assigned Probability	1/3	1/3	1/6	

a. Supply the missing entry in the table.

b. Find $P(A)$ and $P(AB)$.

c. Find $P(A|B)$ and $P(A|C)$.

d. Find $P(A \cup B)$ and $P(A \cup C)$.

e. Are A and B mutually exclusive? Independent?

7. An envelope of seeds contains three nonviable seeds and five viable ones. Considering the eight seeds to be distinguishable:

a. How many different samples of size three can be formed?

b. How many of these samples of size three comprise two viable seeds and one nonviable seed?

c. If a sample of size three is selected at random from this envelope, what is the probability that two of these seeds will be viable and the other nonviable?

8. A random sample of size five is drawn from a large production lot with fraction defective 10%. The probability that this sample will contain no defectives is .59. What is the probability that this sample will contain at least one defective?

9. A factory operates an eight-hour day shift. Five machines of a certain type are used. If one of these machines breaks down, it is set aside and repaired by a crew operating at night. Suppose the probability that a given machine suffers a breakdown during a day's operation is $1/5$.
   a. What is the probability that no machine breakdown will occur on a given day?
   b. What is the probability that two or more machine breakdowns will occur on a given day?

10. A certain virus disease afflicted the families in three adjacent houses in a row of twelve houses. If three houses were randomly chosen from a row of twelve houses, what is the probability that the three houses would be adjacent? Is there reason to conclude that this virus disease is contagious?

11. A geologist, assessing a given tract of land for its oil content, initially rates the land as having
   i.   no oil, with probability 0.7
   ii. 500,000 barrels, with probability 0.2
   iii. 1,000,000 barrels, with probability 0.1.

However, the potential buyer ordered that seismic drillings be performed and found the readings to be "high" based on a "low, medium, high" rating scale. The conditional probabilities, $P(E|H)$, are given in the following table:

H_i \ E_i	Seismic Readings		
	E_1 Low	E_2 Medium	E_3 High
H_1: no oil	.50	.30	.20
H_2: 500,000 bbl.	.40	.40	.20
H_3: 1,000,000 bbl.	.10	.50	.40

a. Find $P(H_1|E_3)$, $P(H_2|E_3)$, and $P(H_3|E_3)$.
b. Suppose the seismic readings had been low. Now find $P(H_1|E_1)$, $P(H_2|E_1)$, and $P(H_3|E_1)$.

12. A manufacturer buys parts from a supplier in lots of 10,000 items. The fraction defective in a lot is usually about 0.1%. Occasionally, a malfunction in the supplier's machinery causes the fraction defective to jump to 3%. Records indicate that the probability of receiving a lot with 3% defective is .1. To check the quality of the supplier's lot, the manufacturer selects a random sample of 200 parts from the lot and observes 3 defectives.

   If the probability of observing 3 defectives when the fraction defective is 0.1% is approximated as .0011 and the probability of observing 3 defectives when the fraction defective is 3% is approximated as .0892, find:

a. the probability that the percentage defective is 0.1%, given 3 defectives are observed in the sample.
b. the probability that the percentage defective is 3%, given 3 defectives are observed in the sample.
c. Based on your answers to a. and b., what would you conclude to be the fraction defective in the lot?

Chapter 5

RANDOM VARIABLES AND
PROBABILITY DISTRIBUTIONS

5.1 Random Variables

Most experiments result in numerical outcomes or events. The outcome
itself may be a numerical quantity such as height, weight, or time, or some
rank ordering of a response. When categorical observations are made such as
good or defective, color of eyes, income bracket, and so on, we are usually
concerned with the number of observations falling into a specified category.
When we observe the outcome of an experiment and assign a numerical
value to that event, we are in fact defining a functional relationship or a cor-
respondence between the events and numerical values.

> A *random variable* is a numerical valued function defined on the
> sample space.

Suppose that y is a random variable. The phrase "defined on a sample space"
means that y takes values associated with sample points that are outcomes of
an experiment. One and only one value of the random variable y is associated
with each point in the sample space. Hence y is a random or chance variable
since it takes values according to some probabilistic model. The values that
the random variable y may assume form one set and the sample points
another. Therefore, the random variable y is said to be a numerical valued
function.

Suppose that a sample of 100 people was randomly drawn from a population of voters and the number favoring candidate Jones was recorded. This
process defines an experiment. The number of voters in the sample favoring
candidate Jones is an example of a _____ _____.   random variable

Further, suppose that of the 100 voters in the sample, 60 favored Jones.
This would not necessarily imply that Jones will win because one could
obtain 60 or more in the *sample* favorable to Jones even though only half of
the voting *population* favor him. In fact, the crucial question is, "What is the
probability that 60 or more of the 100 voters in the sample are favorable to
Jones when actually just 50% of the voting population will vote for him?"
To answer this question, we need to investigate the probabilistic behavior of
the random variable y, the number of favorable voters in a sample of 100

voters. The set of values that the random variable y may assume and the probability, $p(y)$, associated with each value of y define a probability distribution. Hence, before we can use a random variable to make inferences about a population, we must study some basic characteristics of probability distributions.

5.2 Types of Random Variables

Random variables are divided into two classes according to the values that the random variable can assume. If a random variable y can take on only a finite or a countable infinity of distinct values, it is classified as a *discrete random variable*. If a random variable y can take on all of the values associated with the points on a line interval, then y is called a *continuous random variable*. It is necessary to make the above distinction between the discrete and continuous cases because the probability distributions require different mathematical treatment. In fact, calculus is a prerequisite to any complete discussion of continuous random variables. Arithmetic and elementary algebra are all we need to develop discrete probability distributions.

The following would be examples of discrete random variables:
a. The number of voters favoring a political candidate in a given precinct.
b. The number of defective bulbs in a package of twenty bulbs.
c. The number of errors in an income tax return.

Notice that discrete random variables are basically counts and the phrase "the number of" can be used to identify a discrete random variable. The following would be examples of continuous random variables:
a. The time required to complete a medical operation.
b. The height of an experimental strain of corn.
c. The amount of ore produced by a given mining operation.

Classify the following random variables as discrete or continuous:

discrete

a. The number of psychological subjects responding to stimuli in a group of thirty. _____

discrete

b. The number of building permits issued in a community during a given month. _____

discrete

c. The number of amoebae in one cubic centimeter of water. _____

continuous

d. The juice content of six Valencia oranges. _____

continuous

e. The time to failure for an electronic system. _____

continuous

f. The amount of radioactive iodine excreted by rats in a medical experiment. _____

discrete

g. The number of defects in one square yard of carpeting. _____

5.3 Probability Distributions for Discrete Random Variables

The probability distribution for a discrete random variable y consists of the pairs $(y, p(y))$ where y is one of the possible values of the random variable y and $p(y)$ is its corresponding probability. This probability distribution must satisfy two requirements:

1. $\sum\limits_{y} p(y) = $ _____ .

            1

2. _____ $\leqslant p(y) \leqslant$ _____ .

            0; 1

One might express the probability distribution for a discrete random variable y in any one of three ways:

    1. By listing opposite each possible value of y its probability $p(y)$ in a table.
    2. Graphically as a probability histogram.
    3. By supplying a formula together with a list of the possible values of y.

Example 5.1

A businessman has decided to invest $10,000 in each of three common stocks. Four stocks, call them A, B, C, and D, have been recommended to him by a broker and he plans to select three of the four to form an investment portfolio. Unknown to the businessman, stocks A, B, and C will rise in the near future but D will suffer a severe drop in price. If the selection is made at random, find the probability distribution for y, the number of good stocks in the investment portfolio.

Solution

The experiment consists of selecting three of the four stocks. Each of the C_3^4 distinctly different combinations possesses the same chance for selection and these _____ simple events form the sample space. The sample points associated with the four combinations along with their probabilities are shown below. The value $y = 3$ is assigned to E_1 because this combination includes all three of the good stocks. Assign values of y to the other three sample points.

            four

Sample Points	$P(E_i)$	y
E_1: ABC	1/4	3
E_2: ABD	1/4	_____
E_3: ACD	1/4	_____
E_4: BCD	1/4	_____

            2
            2
            2

1. The probability distribution presented as a table is shown below. When $y = 2$, $p(y)$ is $p(2) = P(E_2) + P(E_3) + P(E_4) = $ _____ . Similarly, $p(3) = P(E_1) = $ _____ .

            3/4
            1/4

y	$p(y)$
2	3/4
3	1/4

Note that the two requirements placed on a discrete probability distribution are satisfied in this example. These are

a. _____ .

            $\sum\limits_{y} p(y) = 1$

b. _____ .

            $0 \leqslant p(y) \leqslant 1$

2. The probability distribution can also be graphically presented as a probability histogram.

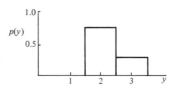

3. A formula appropriate for this probability distribution is

$$p(y) = \frac{C_y^3 \, C_{3-y}^1}{C_3^4} \qquad y = 2, 3.$$

The student may verify that the formula does indeed give the correct values for $p(2)$ and $p(3)$.

4. If this experiment were repeated many times, approximately what

3/4   fraction of the outcomes would result in $y = 2$? _____

What fraction of the total area under the probability histogram lies over

3/4   the interval associated with $y = 2$? _____

Note that the probability distribution provides a theoretical frequency distribution for the hypothetical population associated with the businessman's experiment and thereby relates directly to the content of Chapter 3 in your text.

Example 5.2

A psychological recognition experiment required a subject to classify a set of objects according to whether they had or had not been previously observed. Suppose that a subject can correctly classify each object with probability $p = .7$, that sequential classifications are independent events, and that he is presented with $n = 3$ objects to classify. We are interested in y, the number of correct classifications for the three objects.

Solution

1. This experiment is analogous to tossing three unbalanced coins where correctly classifying an object corresponds to the observation of a head in the toss of a single coin. Each classification results in one of two outcomes, correct or incorrect. Therefore, applying the *mn* rule of Chapter 4,

8   the total number of sample points in the sample space is _____.

2. Let *IIC* represent the sample point for which the classification of the first and second objects is incorrect and the third is correct. Complete the listing of all the sample points in the sample space.

E_1: *III*	E_5: *CII*

CCI E_2: *IIC* E_6: _____

CIC E_3: *ICI* E_7: _____

CCC E_4: *ICC* E_8: _____

3. The sample point E_2 is an *intersection* of three independent events. Applying the Multiplicative Law of Probability,

$$P(E_2) = P(IIC) = P(I)P(I)P(C) = (.3)(.3)(.7) = .063.$$

Similarly, $P(E_1) = .027$ and $P(E_3) = .063$. Calculate the probabilities for all the sample points in the sample space.

$P(E_1) = .027$	$P(E_5) = $ _____		.063
$P(E_2) = .063$	$P(E_6) = $ _____		.147
$P(E_3) = .063$	$P(E_7) = $ _____		.147
$P(E_4) = .147$	$P(E_8) = $ _____ .		.343

4. The random variable y, the number of correct classifications for the set of three objects, takes the value $y = 1$ for sample point E_2. Similarly, we would assign the value $y = 0$ to E_1. Assign a value of y to each sample point in the sample space.

Sample Points	y
E_1	0
E_2, E_3, E_5	1
E_4, E_6, E_7	_____
E_8	_____

2
3

5. The numerical event $y = 0$ contains only the sample point E_1. Summing the probabilities of the sample points in the event $y = 0$, we have $P[y = 0] = P(E_1) = .027$. Similarly, the numerical event $y = 1$ contains three sample points. Summing the probabilities of these sample points, we have $P[y = 1] = p(1) = .189$.
6. The probability distribution $p(y)$ is presented in tabular form below.

y	$p(y)$
0	.027
1	_____
2	_____
3	_____

.189
.441
.343

Calculate the probabilities $p(2)$ and $p(3)$ and complete the table.
7. Present $p(y)$ graphically in the form of a probability histogram.

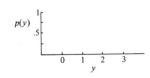

8. After having studied Chapter 6, the student will be able to express this probability distribution in the form of a formula.

Example 5.3

1. Construct a probability table expressing the probability distribution which is given by the formula

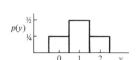

0; ¼
1; ½
2; ¼

$$p(y) = \frac{1}{4} C_y^2 \quad \text{for } y = 0, 1, 2.$$

y	p(y)

2. Construct a probability histogram for the probability distribution in 1.

3. If the experiment implied in 1. were repeated over and over again a large number of times, the frequency histogram would resemble the probability distribution for the random variable y.

Self-Correcting Exercises 5A

1. A subject is shown four photographs, *A, B, C, D*, of crimes that have been committed and asked to select what he considers the two worst crimes. Although all four crimes shown are robberies involving about the same amount of money, *A* and *B* are pictures of robberies committed against private citizens while *C* and *D* are robberies committed against corporations. Assuming that the subject does not show discrimination in his selection, find the probability distribution for *y*, the number of crimes chosen involving private citizens.

2. Someone claims that the following is the probability distribution for a random variable *y*:

y	-1	0	1	2
p(y)	1/10	-2/10	5/10	3/10

Give two reasons why this is not a valid probability distribution.

3. Five equally qualified applicants for a teaching position were ranked in order of preference by the superintendent of schools. If two of the applicants hold master's degrees in education, find the probability distribution for *y*, the number of applicants holding master's in education ranked first or second.

5.4 Continuous Random Variables and Their Probability Distributions

continuous

A _____ random variable can assume a noncountable infinity of values corresponding to points on a line interval. Since the mathematical treatment of continuous random variables requires the use of calculus, we will do no mathematics here, but merely state some basic concepts. The probability distribution for a continuous random variable can be thought of as the limiting histogram for a very large set of measurements utilizing the smallest possible interval width. In such a case, the outline of the histogram would appear as a continuous curve.

Let us illustrate what happens if we begin with a histogram and allow the interval width to get smaller and smaller while the number of measurements gets larger and larger.

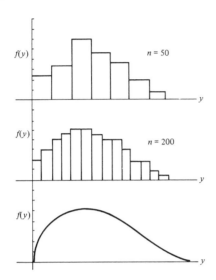

The mathematical function $f(y)$ that traces this curve with varying values of y is called the _____ distribution or the probability density for the random variable y. In the same way that the area under a relative frequency histogram is _____, the area under the curve $f(y)$ is also equal to _____. The area under the curve between two points, say a and b, represents the _____ that the random variable y will fall into the interval from a to b.

probability

one

one

probability

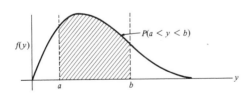

When choosing a model to describe the population of measurements of interest, we must choose $f(y)$ appropriate to our data. Any inferences which we may make will only be as valid as the model we are using. It is therefore very important to know as much as possible about the phenomenon under study that will give rise to the measurements that we record.

The idea of modeling the responses that we will record for an experiment may seem strange to you, but it is readily apparent that you have seen models before, though in a different context. For example, when a physicist says, "The distance, s, traversed by a free-falling body is equal to one-half the force of gravity (g) multiplied by the time (t) squared," and writes $s = \frac{1}{2}gt^2$, he is

modeling
approximations

merely _____ a physical phenomenon with a mathematical formula. These mathematical models merely provide _____ to reality which further need to be verified by experimental techniques.

5.5 Mathematical Expectation

When we develop a probability distribution for a random variable, we are actually proposing a model that will describe the behavior of the random variable in repeated trials of an experiment. For example, when we propose the model for describing the distribution of y, the number of heads in the toss of 2 fair coins, given by

y	$p(y)$
0	1/4
1	1/2
2	1/4

we mean that if the two coins were tossed a large number of times, about one-fourth of the outcomes would result in the outcome "zero heads," one-half would result in the outcome "one head," and the remaining fourth would result in "two heads." A probability distribution is not only a measure of belief that a specific outcome will occur on a single trial, but more important, it actually describes a population of observations on the random variable y. It is reasonable then to talk about and calculate the mean and the standard deviation of a random variable by using the probability distribution as a population model.

The expected value of a random quantity is its average value in the population. In particular, the expected value of y is simply the population mean. The expected value of $(y - \mu)^2$ describes the population variance.

The mean or expected value, $E(y)$, of a discrete random variable y is $\mu = E(y) = \sum\limits_{y} y\, p(y)$. The variance, σ^2 is given as

$$\sigma^2 = E(y - \mu)^2 = \sum\limits_{y} (y - \mu)^2\, p(y).$$

Example 5.4
The probability distribution for y, the number of correct classifications in a psychological experiment, is

y	$p(y)$
0	.15
1	.35
2	.20
3	.30

Find the mean and the standard deviation of the number of correct classifications.

Solution
Before calculating the mean and variance of y, we see that this (is, is not) a
valid probability distribution since

$$\sum_y p(y) = \underline{\hspace{1.5cm}} \quad \text{and} \quad \underline{\hspace{1.5cm}} \leq p(y) \leq \underline{\hspace{1.5cm}}.$$

is

1; 0; 1

1. The expected number of correct classifications is calculated as

$$\mu \doteq E(y) = \sum_y y\, p(y)$$

$$= 0(.15) + 1(.35) + 2(.20) + 3(.30)$$

$$= \underline{\hspace{1.5cm}}.$$

1.65

2. To calculate the variance, σ^2, we can use the definition formula given
   above *or* its equivalent computational form given as

$$\sigma^2 = E(y^2) - \mu^2.$$

Using the computational formula, we need

$$E(y^2) = 0^2(.15) + 1^2(.35) + 2^2(.20) + 3^2(.30)$$

$$= \underline{\hspace{1.5cm}}.$$

3.85

Then $\quad \sigma^2 = 3.85 - (1.65)^2 = \underline{\hspace{1.5cm}}$

1.1275

and $\quad \sigma = \underline{\hspace{1.5cm}}.$

1.06

Example 5.5
Construct the probability histogram for the distribution of the number of
correct classifications given in Example 5.4. Visually locate the mean and
compare it with the computed value, $\mu = 1.65$.

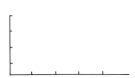

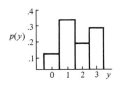

The approximate value of the mean is \underline{\hspace{1.5cm}}. Its value is close to that
calculated and provides an easy check on the calculated value of $E(y)$.

1.5

Example 5.6
A corporation has four investment possibilities: *A*, *B*, and *C* with respective
gains of 10, 20, and 50 million dollars and investment *D* with a loss of 30
million dollars. If one investment will be made and the probabilities of
choosing *A*, *B*, *C*, or *D* are .1, .4, .2, and .3, respectively, find the expected
gain for the corporation.

Solution
The random variable is y, the corporation's gain, with possible values 10, 20, 50, and –30 million dollars. The probability distribution for y is given as

y (in millions)	$p(y)$
10	.1
20	.4
50	.2
–30	.3

10

The expected gain, $E(y) =$ _____ million dollars.

Example 5.7
A parcel post service which insures packages against loss up to $200 wishes to reevaluate its insurance rates. If one in every thousand packages had been reported lost during the last several years, what rate should be charged on a package insured for $200 so that the postal service's expected gain is zero? Administrative costs will be added to this rate.

Solution
Let y be the gain to the parcel post service and let r be the rate charged for insuring a package for $200.
1. Complete the probability distribution for y.

y	$p(y)$
r	.999
_____	.001

-(200 – r)

2. If $E(y)$ is to be zero, we need to solve the equation

$$\sum_y y\, p(y) = 0.$$

Hence for our problem

$$r\,(.999) - (200 - r)\,(.001) = 0$$

$.20

$r =$ _____ .

Self-Correcting Exercises 5B

1. Let y be a discrete random variable with probability distribution given as

y	–2	–1	0	1
$p(y)$	1/9	1/9	4/9	

a. Find $p(1)$.

b. Find $\mu = E(y)$.

c. Find σ, the standard deviation of y.

2. A police car visits a given neighborhood a random number of times, y, per evening. If $p(y)$ is given by

y	$p(y)$
0	.1
1	.6
2	.2
3	.1

a. Find $E(y)$.

b. Find σ^2.

3. Refer to Exercise 2. What is the probability that the patrol car will visit the neighborhood at least twice in a given evening?

4. You are given the following information. An insurance company wants to insure a $30,000 home against fire. One in every hundred of such homes is likely to have a fire; 75% of the homes having fires will suffer damages amounting to $15,000 while the remaining 25% will suffer total loss. Ignoring all other partial losses, what premium should the company charge in order to break even?

EXERCISES

1. A car rental agency has three Fords and two Chevrolets left in its car pool. If two cars are needed and the keys are randomly selected from the keyboard, find the probability distribution for y, the number of Fords in the selection.

2. An experiment is run in the following manner. The colors red, yellow, and blue are each flashed on a screen for a short period of time. A subject views the colors and is asked to choose the one which he feels was flashed for the longest amount of time. The experiment is repeated 3 times with the same subject.

   a. If all the colors were flashed for the same length of time, give the probability distribution for y, the number of times the subject chose the color red. Assume that his three choices are independent.

   b. Construct a probability histogram for $p(y)$ found in part a.

3. A publishing company is considering the introduction of a monthly gardening magazine. Advance surveys show the initial market for the magazine will be approximated by the following distribution for y, the number of subscribers.

y	$p(y)$
5,000	.30
10,000	.35
15,000	.20
20,000	.10
25,000	.05

Find the expected number of subscribers and the standard deviation of the number of subscribers.

4. Refer to Exercise 3. Suppose the company expects to charge $10 for an annual subscription. Find the mean and standard deviation of the revenue the company can expect from the annual subscriptions of the initial subscribers.

5. Refer to Exercise 4. Production and distribution costs for the gardening magazine are expected to amount to slightly over $100,000. What is the probability that revenue from initial subscriptions will fail to cover these costs?

6. The following is the probability function for a discrete random variable y:

$$p(y) = (.1)(y + 1) \qquad y = 0, 1, 2, 3.$$

a. Find $\mu = E(y)$ and σ^2.
b. Construct a probability histogram for $p(y)$.

7. The probability of hitting oil in a single drilling operation is $1/4$. If drillings represent independent events, find the probability distribution for y, the number of drillings until the first success ($y = 1, 2, 3, \ldots$). Proceed as follows:

a. Find $p(1)$.
b. Find $p(2)$.
c. Find $p(3)$.
d. Give a formula for $p(y)$.
Note that y can become infinitely large.

e. Will $\displaystyle\sum_{y=1}^{\infty} p(y) = 1$?

8. History has shown that buildings of a certain type of construction suffer fire damage during a given year with probability .01. If a building suffers fire damage, it will result in either a 50% or a 100% loss with probabilities .7 and .3, respectively. Find the premium required per $1,000 coverage in order that the expected gain for the insurance company will equal zero (break-even point).

9. Experience has shown that a rare disease will cause partial disability with probability .6, complete disability with probability .3, and no disability with probability .1. Only one in ten thousand will become afflicted with the disease in a given year. If an insurance policy pays $20,000 for partial disability and $50,000 for complete disability, what premium should be charged in order that the insurance company break even (that is, in order that the expected loss to the insurance company will be zero)?

Chapter 6

THE BINOMIAL PROBABILITY DISTRIBUTION

6.1 The Binomial Experiment

Having discussed probability and its role in inference making, we now turn our attention to an important discrete random variable called the binomial random variable. The binomial random variable and its probability distribution provide a model that is very useful and has wide application in the social, biological, and physical sciences.

Many experiments in the social, biological, and physical sciences can be reduced to a series of trials resembling the toss of a coin whereby the outcome on each toss will be either a head or a tail. Consider the following analogies:

1. A student answers a multiple-choice question correctly (head) or incorrectly (tail).
2. A voter casts his ballot either for candidate A (head) or against him (tail).
3. A patient having been treated with a particular drug either improves (head) or does not improve (tail).
4. A subject either makes a correct identification (head) or an incorrect one (tail).
5. A licensed driver either has an accident (head) or does not have an accident (tail) during the period his license is valid.
6. An item from a production line is inspected and classified as either defective (head) or not defective (tail).

If any of the above situations were repeated n times and we counted the number of "heads" that occurred in the n trials, the resulting random variable would be a _____ random variable. Let us examine what characteristics these experiments have in common. We shall call a head a success (S) and a tail a failure (F). Note well that a success does not necessarily denote a desirable outcome, but rather identifies the event of interest.

The five defining characteristics of a binomial experiment are:

1. The experiment consists of n identical trials.
2. Each trial results in one of two outcomes, success (S) or failure (F).

binomial

3. The probability of success on a single trial is equal to p and remains constant from trial to trial. The probability of failure is $q = 1 - p$.
4. The trials are independent.
5. Attention is directed to the random variable y, the total number of successes observed during the n trials.

Although very few real life situations perfectly satisfy all five characteristics, this model can be used with fairly good results if the violations are "moderate." The next several examples will illustrate binomial experiments.

Example 6.1
A procedure (the "triangle test") often used to control the quality of name-brand food products utilizes a panel of n "tasters." Each member of the panel is presented three specimens, two of which are from batches of the product known to possess the desired taste while the other is a specimen from the latest batch. Each panelist is asked to select the specimen which is different from the other two. If the latest batch does possess the desired taste, then the probability that a given taster will be "successful" in selecting the specimen from the latest batch is _____ . If there is no communication among the panelists their responses will comprise n independent _____ with probability of success on a given trial equal to _____ .

1/3
trials
1/3

Example 6.2
Almost all auditing of accounts is done on a sampling basis. Thus, an auditor might check a random sample of n items from a ledger or inventory list comprising a large number of items. If 1% of the items in the ledger are erroneous, then the number of erroneous items in the sample is essentially a _____ random variable with n trials and probability of success (finding an erroneous item) on a given trial equal to _____ .

binomial
.01

Example 6.3
No treatment has been known for a certain serious disease for which the mortality rate in the United States is 70%. If a random selection is made of 100 past victims of this disease in the United States, the number, y_1, of those in the sample who died of the disease is essentially a binomial random variable with $n = $ _____ and $p = $ _____ . More important, if observation is made of the next 100 persons in the United States who will in the future become victims of this disease, the number, y_2, of these who will die from the disease has a distribution approximately the same as that of y_1 if conditions affecting this disease remain essentially constant for the time period considered.

100; .70

Example 6.4
The continued operation (reliability) of a complex assembly often depends on the joint survival of all or nearly all of a number of similar components. Thus, a radio may give at least 100 hours of continuous service if no more than two of its ten transistors fail during the first 100 hours of operation. If the ten transistors in a given radio were selected at random from a large lot of transistors, then each of these (ten) transistors would have the same probabil-

ity, p, of failing within 100 hours, and the number of transistors in the radio which will fail within 100 hours is a _____ random variable with _____ trials and probability of success on each trial equal to _____. (Success is a word that denotes one of the two outcomes of a single trial and does not necessarily represent a desired outcome.)

Three experiments are described below. In each case state whether or not the experiment is a binomial experiment. If the experiment is binomial, specify the number, n, of trials and the probability, p, of success on a given trial. If the experiment is not binomial, state which characteristics of a binomial experiment are not met.

1. A fair coin is tossed until a head appears. The number of tosses, y, is observed. If binomial, $n =$ _____ and $p =$ _____. If not binomial, list characteristic(s) (1, 2, 3, 4, and 5) violated. _____
2. The probability that an applicant scores above the 90th percentile on a qualifying examination is .10. The examiner is interested in y, the number of applicants out of 25 taking the examination that score above the 90th percentile. If binomial, $n =$ _____ and $p =$ _____. If not binomial, list characteristic(s) (1, 2, 3, 4, and 5) violated. _____
3. A sample of five transistors will be selected at random from a box of twenty transistors of which ten are defective. The experimenter will observe the number, y, of defective transistors appearing in the sample. If binomial, $n =$ _____ and $p =$ _____. If not binomial, list characteristic(s) (1, 2, 3, 4, and 5) violated. _____

binomial
10
p

not binomial
1, 5

25; .10
none

not binomial
3, 4

6.2 The Binomial Theorem

The following identity is proved in most high school algebra books:

$$(a + b)^n = \sum_{y=0}^{n} \frac{n!}{y!\,(n-y)!} a^y b^{n-y}$$

where a and b are any real numbers and n is a positive integer. This identity is known as the binomial theorem. The coefficients in the sum are known as binomial coefficients and we can write

$$\frac{n!}{y!\,(n-y)!} = C_y^n$$

to represent the coefficient of $a^y b^{n-y}$ in the expansion of $(a + b)^n$. For the special case when $n = 2$ we find

$$(a + b)^2 = \sum_{y=0}^{2} C_y^2 a^y b^{2-y}$$

$$= a^2 + 2ab + b^2.$$

When $n = 3$, we have

$$(a + b)^3 = \sum_{y=0}^{3} C_y^3 \, a^y \, b^{3-y}$$

$$= a^3 + 3a^2b + 3ab^2 + b^3.$$

If a is replaced by a probability, p, and b is replaced by the probability, $q = 1 - p$, then the binomial theorem implies that

$$\sum_{y=0}^{n} C_y^n \, p^y \, q^{n-y} = (p + q)^n.$$

Note the following points about this binomial identity involving p and q when $0 < p < 1$:

1. The terms $C_y^n \, p^y \, q^{n-y}$ are each positive.
2. The sum of the terms is $(p + q)^n = 1$.
3. Since the total can be no greater than any of its parts

$$0 \leqslant C_y^n \, p^y \, q^{n-y} \leqslant 1.$$

Thus the function

$$p(y) = C_y^n \, p^y \, q^{n-y}, \qquad y = 0, 1, 2, \ldots, n$$

satisfies the two requirements of a probability function for a discrete random variable; namely,

$$1. \; 0 \leqslant p(y) \leqslant 1 \quad \text{and} \quad 2. \; \sum_y p(y) = 1.$$

6.3 The Binomial Probability Distribution

The probability distribution for y, the number of successes in n trials where p is the probability of success on a given trial is given by the formula

$$\left[\; p(y) = C_y^n \, p^y \, q^{n-y} \; \right]$$

for the values $y = 0, 1, 2, \ldots, n$ with $q = 1 - p$. To help us in using the formula for the binomial distribution, consider the following situation:

Example 6.5
The president of an agency specializing in public opinion surveys claims that approximately 70% of all people to whom they send questionnaires respond by filling out and returning the questionnaire. Four such questionnaires are sent out. Let y be the number of questionnaires which are filled out and

returned. Then y is a binomial random variable with $n =$ _____ and
$p =$ _____.

1. The probability that no questionnaires are filled out and returned is

$$p(0) = C_0^4 (.7)^0 (.3)^4 = \frac{4!}{0!\,4!} (.7)^0 (.3)^4$$

$$= (.3)^4 = \underline{\hspace{2cm}}.$$

.0081

2. The probability that exactly three questionnaires are filled out and returned is

$$p(3) = C_1^4 (.7)^3 (.3)^1$$

$$= 4 (.343) (.3) = \underline{\hspace{2cm}}.$$

.4116

3. The probability that at least three questionnaires are filled out and returned is

$$P[y \geqslant 3] = p(3) + p(4)$$

$$= p(3) + C_4^4 (.7)^4 (.3)^0$$

$$= .4116 + .2401$$

$$= .6517.$$

Example 6.6

A marketing research survey shows that approximately 80% of the car owners surveyed indicated that their next car purchase would be either a compact or an economy car. If the 80% figure is taken to be correct, and five prospective buyers are interviewed,

a. Find the probability that all five indicate that their next car purchase would be either a compact or an economy car.
b. Find the probability that at most one indicates that his next purchase will be either a compact or an economy car.

Solution

Let y be the number of car owners who indicate that their next purchase will be a compact or an economy car. Then $n =$ _____ and $p =$ _____ and the distribution for y is given by

5; .8

$$p(y) = C_y^5 (.8)^y (.2)^{5-y} \qquad y = 0, 1, \ldots, 5.$$

a. The required probability is $p(5)$ which is given by

$$p(5) = C_5^5 (.8)^5 (.2)^0$$

$$= (.8)^5$$

$$= .32768.$$

b. The probability that at most one car owner indicates that his next purchase will be either a compact or an economy car will be

$$P[y \leqslant 1] = p(0) + p(1).$$

For $y = 0$,

$$p(0) = C_0^5 (.8)^0 (.2)^5$$

$$= (.2)^5$$

.00032

$$= \underline{\qquad}.$$

For $y = 1$,

$$p(1) = C_1^5 (.8)^1 (.2)^4$$

$$= 5 (.8) (.0016)$$

.0064

$$= \underline{\qquad}.$$

Hence $P[y \leqslant 1] = .0064 + .00032$

.00672

$$= \underline{\qquad}.$$

As you might expect, the calculation of the binomial probabilities becomes quite tiresome as the number of trials increases. Table 1 of binomial probabilities, Appendix II in your text, can be used to find binomial probabilities for values of $p = .01, .05, .10, .20, \ldots, .90, .95, .99$ when $n = 5, 10, 15, 20, 25$.
1. The tabled entries are not the individual terms for binomial probabilities, but rather cumulative sums of probabilities, beginning with $y = 0$ up to and including the value $y = a$. By formula, the entries for n, p, and a are

$$\left[\sum_{y=0}^{y=a} p(y) = p(0) + p(1) + \ldots + p(a) \right]$$

2. By using a tabled entry, which is $\sum_{y=0}^{a} p(y)$, these tables allow the user to find
   a. left-tail cumulative sums,

$$P[y \leqslant a] = \sum_{y=0}^{a} p(y),$$

   b. right-tail cumulative sums,

$$P[y \geqslant a] = 1 - \sum_{y=0}^{a-1} p(y),$$

c. or individual terms such as

$$P[y = a] = \sum_{y=0}^{a} p(y) - \sum_{y=0}^{a-1} p(y).$$

Example 6.7

Refer to Example 6.6. Let us find the probabilities asked for by using Table 1.

Solution

For this problem, we shall use the table $n = 5$ and $p = .8$.

1. To find the probability that $y = 5$ we proceed as follows.

$$p(5) = [p(0) + p(1) + p(2) + p(3) + p(4) + p(5)]$$

$$- [p(0) + p(1) + p(2) + p(3) + p(4)]$$

$$= \sum_{y=0}^{5} p(y) - \sum_{y=0}^{4} p(y)$$

$$= 1 - .672$$

$$= .328.$$

2. To find the probability that $y \leqslant 1$, we need

$$P[y \leqslant 1] = p(0) + p(1)$$

$$= \sum_{y=0}^{1} p(y)$$

$$= \underline{\hspace{2cm}}.$$

.007

3. Let us extend the problem and find the probabilities associated with the terms, $y = 2$ and $y = 3$.

For $y = 2$,

$$p(2) = \sum_{y=0}^{2} p(y) - \sum_{y=0}^{1} p(y)$$

$$= .058 - .007$$

$$= \underline{\hspace{2cm}}.$$

.051

For $y = 3$,

$$p(3) = \sum_{y=0}^{3} p(y) - \sum_{y=0}^{2} p(y)$$

$$= .263 - .058$$

.205

$$= \underline{\hspace{2cm}}.$$

4. Complete the following table:

y	$p(y)$
0	_____
1	_____
2	_____
3	_____
4	_____
5	_____

.000
.007
.051
.205
.409
.328

1

with $\displaystyle\sum_{y=0}^{5} p(y) = \underline{\hspace{2cm}}.$

5. Graph this distribution as a probability histogram.

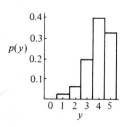

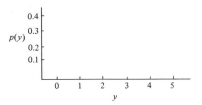

Example 6.8
Using Table 1, find the probability distribution for y if $n = 5$ and $p = \frac{1}{2}$, and graph the resulting probability histogram.

Solution
1. To find the individual probabilities for $y = 0, 1, 2, \ldots, 5$, we need but subtract successive entries in the table for $n = 5, p = .5$.

.031

$$p(0) = \sum_{y=0}^{0} p(y) = \underline{\hspace{2cm}}$$

.157

$$p(1) = \sum_{y=0}^{1} p(y) - \sum_{y=0}^{0} p(y) = .188 - .031 = \underline{\hspace{2cm}}$$

$p(2) = \sum_{y=0}^{2} p(y) - \sum_{y=0}^{1} p(y) = .500 - .188 =$ _____ .312

$p(3) =$ _____ $- .500 =$ _____ .812; .312

$p(4) =$ _____ $- .812 =$ _____ .969; .157

$p(5) = 1.000 -$ _____ $=$ _____ . .969; .031

2. Using the results of 1, we find the probability histogram to be symmetric about the value $y =$ _____ .　　　　2.5

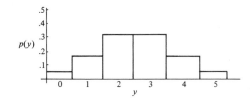

Example 6.9
Find the probability distribution for y if $n = 5$ and $p = .3$, and graph the probability histogram in this case.

Solution
1. Again, subtracting successive entries for $n = 5, p = .3$, we have

$p(0) =$ _____ .168

$p(1) = .528 - .168 =$ _____ .360

$p(2) = .837 - .528 =$ _____ .309

$p(3) = .969 - .837 =$ _____ .132

$p(4) = .998 - .969 =$ _____ .029

$p(5) = 1 - .998 =$ _____ . .002

2. Graph the resulting histogram.

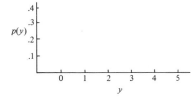

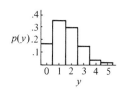

symmetric
right

left

p

In comparing the histogram in Examples 6.7, 6.8, and 6.9, notice that when $p = \frac{1}{2}$, the histogram is _____ . However, if $p = .8$, which is greater than $\frac{1}{2}$, the mass of the probability moves to the _____ with p; and for $p = .3$, which is less than $\frac{1}{2}$, the mass of the probability distribution moves to the _____ with p. Locating the center of the distribution by eye, we see that the mean of the binomial distribution varies directly as _____ , the probability of success.

Let us consider two more examples. You are now free either to calculate the probabilities by hand or to use the tables when appropriate.

Example 6.10
To test two methods of presentation of a given subject, ten pairs of identical twins were selected and one from each pair randomly assigned to method A, the other to method B. At the end of the teaching periods, all the twins were given the same test and their grades were recorded. Let y be the number of twins taught using method A who did better on the exam than their respective twin taught by method B.
A. If there is really no difference between the two methods, the probability that twin A does better than twin B on the exam can be taken to be $p = .5$. What is the probability that 8 or more twins in the A group did better than their "B twins"?

Solution
Since $n = 10$ and $p = .5$, we can use the table to find $P[y \geqslant 8]$.

$$P[y \geqslant 8] = \sum_{y=8}^{10} p(y)$$

$$= 1 - \sum_{y=0}^{7} p(y)$$

.945

$$= 1 - \underline{\hspace{2cm}}$$

.055

$$= \underline{\hspace{2cm}} .$$

B. If $p = \frac{1}{2}$, what is the probability that 6 or more of the B twins had better scores than their A twins?

Solution
Since the random variable y is the number of A twins excelling, then $y \leqslant 4$ when 6 or more B twins excel. Hence,

.377

$$P[y \leqslant 4] = \sum_{y=0}^{4} p(y) = \underline{\hspace{2cm}} .$$

Example 6.11

A preliminary investigation reported that approximately 30% of locally grown poultry were infected with an intestinal parasite which, although not harmful to those consuming the poultry, decreased the usual weight growth rates in the birds and thereby caused a loss in revenue to the growers. A diet supplement believed to be effective against this parasite was added to the birds' rations. During the preparation of poultry that had been fed the supplemental rations for at least two weeks, of 25 birds examined, 3 birds were still found to be infected with the intestinal parasite.

A. If the diet supplement is ineffective, what is the probability of observing 3 or fewer birds infected with the intestinal parasite?

Solution

With $n = 25$ and $p = .3$, we can use the binomial tables in your text to find $P[y \leqslant 3]$.

$$P[y \leqslant 3] = \sum_{y=0}^{3} p(y)$$

$$= \underline{\hspace{2cm}}.$$ 　　　　　　　.033

B. If in fact the diet supplement was effective and reduced the infection rate to 10%, what is the probability of observing 3 or fewer infected birds?

Solution

We can use the same tables with $n = 25$ and $p = $ _____ . Hence, 　　.1

$$P[y \leqslant 3] = \sum_{y=0}^{3} p(y)$$

$$= \underline{\hspace{2cm}}.$$ 　　　　　　　.902

Notice that the sample results are much more probable if the diet supplement (was, was not) effective in reducing the infection rate below 30%. 　　was

Self-Correcting Exercises 6A

1. A city planner claims that 20% of all apartment dwellers move from their apartments within a year from the time they first moved in. In a particular city, 7 apartment dwellers who had given notice of termination to their landlords are to be interviewed.
   a. If the city planner is correct, what is the probability that 2 of the 7 had lived in the apartment for less than one year?
   b. What is the probability that at least 6 had lived in their apartment for at least one year?

2. Suppose that 70% of the first class mail from New York to California is delivered within four days after being mailed. If twenty pieces of first class mail are mailed from New York to California,
   a. find the probability that at least 15 pieces of mail arrive within 4 days of the mailing date.
   b. find the probability that 10 or fewer pieces of mail arrive later than four days after the mailing date.
3. In the past history of a certain serious disease it has been found that about ½ of its victims recover.
   a. Find the probability that exactly four of the next fifteen patients suffering from this disease will recover.
   b. Find the probability that at least four of the next fifteen patients afflicted with this disease will recover.
4. On a certain university campus a student is fined $1.00 for the first parking violation of the academic year. The fine is doubled for each subsequent offense, so that the second violation costs $2.00, the third $4.00, and so on. The probability that a parking violation on a given day is detected is .10. Suppose that a certain student will park illegally on each of 20 days during a given academic year.
   a. What is the probability he will not be fined?
   b. What is the probability that his fines will total no more than $15.00?

6.4   The Mean and Variance for the Binomial Random Variable

The mean and standard deviation of a probability distribution are important summary measures that are used to locate the center of the distribution and to describe the dispersion of the measurements about the mean. We noted in the last section that the mass of the probability distribution for a binomial with $n = 5$ shifted with the value of p. Further calculations with various values for n and p will show that the mean varies directly with p.

The mean and variance (and hence the standard deviation) can be found using the expectation definitions of Chapter 5 together with

$$p(y) = C_y^n \, p^y \, q^{n-y} \qquad y = 0, 1, 2, \ldots, n.$$

It can be shown by those willing to tackle the algebra (and can be *used* by those not so willing) that for a binomial experiment consisting of n trials with the probability of success equal to p,

1. $\mu = E(y) = np,$

2. $\sigma^2 = E(y - \mu)^2 = npq,$

3. $\sigma = \sqrt{npq}.$

Tchebysheff's Theorem can be used in conjunction with the distribution of a binomial random variable since *at least* $(1 - 1/k^2)$ of *any* distribution lies within k standard deviations of the mean. However, when the number of trials, n, becomes large, the Empirical Rule can be used with fairly accurate results. The interval $np \pm 2\sqrt{npq}$ should contain approximately 95% of the distribution while the interval $np \pm 3\sqrt{npq}$ should contain approximately 99.7% of the distribution.

Example 6.12
Suppose it is known that 10% of the citizens of the United States are in favor of increased foreign aid. A random sample of 100 United States citizens are questioned on this issue.
1. Find the mean and standard deviation of y, the number of citizens favoring increased foreign aid.
2. Within what limits would we expect to find the number favoring increased foreign aid?

Solution
With $n = 100$ and $p = .1$,

1. $\mu = np = 100(.1) =$ _____   10

             $\sigma^2 = npq = 100(.1)(.9) =$ _____   9

             $\sigma = \sqrt{npq} = \sqrt{\rule{2em}{0pt}} =$ _____.   9; 3

2. From part 1, $\mu = 10$ and $\sigma = 3$. Using two standard deviations we find the interval $\mu \pm 2\sigma$ to be $10 \pm 2(3)$ or 10 ± 6. With a 95% chance of being correct, we would expect the number of citizens favoring increased foreign aid to lie between _____ and _____ if, in fact, $p = .1$.   4; 16

Example 6.13
A random sample of sixty-four people were asked to state a preference for candidate A or candidate B. If there is no underlying preference for either candidate, then the probability that an individual chooses candidate A will be $p =$ _____.   .5
1. What will be the expected number and standard deviation of preferences for candidate A?
2. Within what limits would you expect the number of stated preferences for candidate A to lie?

Solution
Let y be the number of people stating a preference for candidate A. If there really is no preference for either candidate (that is, the voter selects a candidate at random) then y has a binomial distribution with $n = 64$ and $p =$ _____.   .5

1. $\mu = np = 64(.5) =$ _____   32

             $\sigma^2 = npq = 64(.5)(.5) =$ _____   16

             $\sigma = \sqrt{npq} =$ _____.   4

2. From part 1, $\mu = 32$ and $\sigma = 4$. Hence, $\mu \pm 2\sigma = 32 \pm 8$. With a 95% chance

24; 40

of being correct, we would expect the number of preferences for candidate A to lie between _____ and _____ if, in fact, $p = .5$.

Self-Correcting Exercises 6B

1. Harvard University has found that about 90% of its accepted applicants for enrollment in the freshman class will actually take a place in that class. In 1967, 1,360 applications to Harvard were accepted. Within what limits would you expect to find the size of the freshman class at Harvard in the fall of 1967?

2. If 20% of the registered voters in a given city belong to a minority group and voter registration lists are used in selecting potential jurors, within what limits would you expect the number of minority members on a list of 80 potential jurors to lie if the 80 persons were randomly selected from the voter registration lists?

3. A television network claims that its Wednesday evening prime time program attracts 40% of the television audience. If a random sample of 400 television viewers were asked whether they had seen the previous show, within what limits would you expect the number of viewers who had seen the previous show to lie if the 40% figure is correct? What would you conclude if the interviews revealed that 96 of the 400 had actually seen the previous show?

6.5 Lot Acceptance Sampling

Most manufacturing plants can be thought of as processors that accept raw materials and turn them into finished products. Efficient operation would require that the number of defective items accepted for processing be kept to a minimum and the number of acceptable finished products be kept at a maximum.

These goals can be achieved in different ways. A manufacturer producing television sets would obviously test and adjust *each* set before it leaves the plant, but would probably not test each transistor in an incoming lot before accepting the whole shipment. He would probably accept or reject the shipment depending upon the number of defective transistors observed in a random sample drawn from that lot. Sometimes the act of testing an item is destructive, so that each item cannot be individually tested. Testing whether a flashbulb produces the required intensity of light obviously destroys the flashbulb.

The process of screening lots is an inferential procedure in which a decision about the proportion defective in a lot (population) is made. The sampling of items from incoming or outgoing lots or the sampling of items from a production line closely approximates the defining characteristics of a binomial experiment. Therefore, the number of defectives in a sample of size n will be distributed as a binomial random variable with parameter p, the proportion defective items in the population sampled.

A number of sampling schemes are used in industry, the simplest of which is the following:

From the lot select n items at random. Record the number, y, of defective items found in the sample. If y is less than or equal to a prescribed number, a, accept the lot; otherwise reject the lot. This maximum number, a, of allowable defectives is called the *acceptance number* for the plan.

The above plan is called a single sampling plan. Any such plan is defined by specifying values for the numbers _____ (sample size) and _____ (acceptance number).

   How does one decide whether to use plan A ($n = 10$, $a = 1$) or plan B ($n = 20$, $a = 2$), or some other plan? One acceptable criterion is that the probability of accepting a good lot shall be (high, low) and that the probability of accepting a bad lot shall be _____. Thus we might select plan B (with the larger sample size) rather than plan A if good lots have a higher probability of acceptance and bad lots have a lower probability of acceptance when plan B is used. To obtain this comparison we construct the *operating characteristic* curve for each of these plans. The operating characteristic curve is a graph which shows the probability of acceptance for an incoming lot with fraction defective p. The curve will be shown for values of p ranging from 0 (perfect lot) to 1. If a lot contains no good items ($p =$ _____) then the probability that it will be accepted is _____. If a lot contains no defective items ($p =$ _____) it is certain to be accepted. For intermediate values of p the operating characteristic curves for two different plans will not in general coincide.

n; a

high
low

1
0
0

Example 6.14
Construct an operating characteristic curve for the sampling plan $n = 10$, $a = 1$.

Solution
Using this particular plan, we take a sample of size 10 and accept the lot if no more than 1 defective item is found. We assume that the lot is (large, small) enough to justify treating y, the number of defectives found in the sample, as a _____ random variable.

large

binomial

1. Suppose that the lot contains 10% defective items ($p = .1$). The probability of accepting this lot is

$$P[y \leqslant 1] = p(0) + p(1) = C_0^{10}(.1)^0(.9)^{10} + C_1^{10}(.1)^1(.9)^9.$$

It is not necessary to complete this calculation since the result correct to the nearest thousandth may be read directly from Table 1 in your text. Thus,

$$C_0^{10}(.1)^0(.9)^{10} + C_1^{10}(.1)^1(.9)^9 = \text{_____}.$$

.736

2. By referring to Table 1 (text), obtain the probabilities of acceptance which are omitted in the following table. Complete the table by filling in the missing entries when $n = 10$ and $a = 1$.

Fraction Defective, p	Probability of Lot Acceptance
.01	.996
.05	_____
.10	.736
.20	_____
.30	_____
.40	_____
.50	.011

.914

.376
.149
.046

3. A graph may now be constructed showing the probability of acceptance as a function of the fraction defective in the incoming lot. The curve so obtained is called the *operating characteristic* curve for the sampling plan.

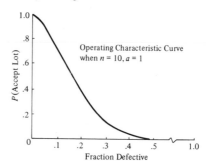

Example 6.15
Construct an O.C. curve for the plan $n = 20$, $a = 2$.

Solution
The probability of accepting a lot under this plan is the probability of obtaining no more than _____ defective items in a random sample of size 20. Thus, the probability of accepting an incoming lot with fraction defective p is

2

$$P[y \leqslant 2] = \sum_{y=\underline{\quad}}^{\overline{2}} C_y^{20} \, p^y (1 - p)^{20-y}$$

(fill in the summation limits)

$\sum\limits_{y=0}^{2}$

1. Using Table 1 (text) complete the following table:

Fraction Defective, p	Probability of Lot Acceptance
.05	_____
.10	_____
.20	_____
.30	_____

.925
.677
.206
.035

2. Complete the O.C. curve by labeling and scaling the following axes, plotting the points obtained from the above table, and joining the points with a smooth curve.

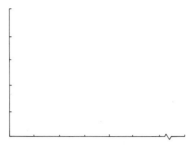

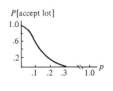

Since an O.C. curve falls as one moves to the right, the probability of accepting a lot containing a high fraction defective is (more, less) than the probability of accepting a good lot.

less

Example 6.16

To aid in the comparison of plan A ($n = 10, a = 1$) and plan B ($n = 20, a = 2$) we shall show their operating characteristic curves on the same graph, using acceptance values recorded in the following table:

Fraction Defective,	Probability of Acceptance	
p	Plan A	Plan B
.00	1.000	1.000
.05	.914	.925
.10	.736	.677
.20	.376	.206
.30	.149	.035
.40	.046	.004
.50	.011	.000

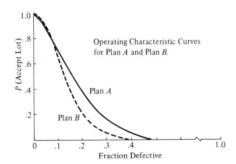

Operating Characteristic Curves for Plan A and Plan B.

The two curves should cross at about $p = .06$. Thus, the probability of accepting a lot with fraction defective less than .06 is (higher, lower) with plan B than with plan A. The probability of accepting a lot with fraction defective more than .06 is _____ with plan B than with plan A. Thus plan B is more sensitive in discriminating between good and bad lots. The expense of inspecting a larger sample (as in plan B) may be justified by the greater sensitivity of plan B as compared with plan A.

higher

lower

Self-Correcting Exercises 6C

1. Large lots of portable radios are accepted in accordance with the sampling plan with sample size $n = 4$ and acceptance number $a = 1$.
   a. Complete the following table and construct the O.C. curve for this plan. The axes should be properly labeled and scaled.

Fraction Defective, p	0	.10	.30	.50	1
Probability of Accepting	_____	.95	_____	.31	_____

   b. State two essentially different ways in which one might modify the above sampling plan to increase the probability of accepting a lot with fraction defective $p = .10$.
2. To discover the effect on acceptance probabilities of varying the sample size we study the additional sampling plans $(n = 10, a = 1)$ and $(n = 25, a = 1)$.
   a. Use Table 1 in your text to complete the following table:

Fraction Defective, p	0	.10	.30	.50	1.0
Probability of Accepting					
$n = 10, a = 1$	_____	.74		.01	_____
$n = 25, a = 1$	_____	_____	.00	_____	_____

   b. Construct the O.C. curves for the plan in Exercise 1 and the plans considered in part a.
   c. If the acceptance number is kept the same and the sample size increased, what is the effect on the probability of accepting a given lot?
3. To discover the effect on acceptance probabilities of varying the acceptance number we study the additional sampling plans $(n = 25, a = 3)$ and $(n = 25, a = 5)$.
   a. Use Table 1 in your text to complete the following table:

Fraction Defective, p	0	.10	.30	.50	1.0
Probability of Accepting					
$n = 25, a = 3$	_____	_____	.03	_____	_____
$n = 25, a = 5$	_____	.97	_____	_____	_____

   b. Construct the O.C. curves for the plans considered in part a. together with the O.C. curve for the plan $(n = 25, a = 1)$ (see Exercise 2) on the same set of axes.
   c. If the sample size is kept the same and the acceptance number increased, what is the effect on the probability of accepting a given lot?

6.6 A Test of an Hypothesis

We have described binomial populations in the first four sections of this chapter using the viewpoint that if we know the values of n and p, the population distribution can be found and the mean and variance of the population can be calculated. In short, we are able to calculate the probability of our sample outcome. Let us now look at the same problem in a different light. Given that we have a sample of n measurements from a dichotomous population with y of these outcomes as "successes," what information about the value of p can be gleaned from the sample? As we have seen, the mass of the probability distribution shifts with the value of _____, the probability of success, so that certain values of y are highly probable for one value of p and highly improbable for other values of p.

   Our approach will be to draw a random sample from a dichotomous population and decide whether we shall accept or reject an hypothesized value for p. The decision will be made on the basis of whether the sample results are highly _____ and support the hypothesized value or are _____ and fail to support the hypothesized value. Our procedure is very similar to a court trial in which the accused is assumed innocent until proved guilty. In fact, our sample acts as the _____ for or against the accused. What we do is to compare the hypothesized value with reality. Let us illustrate how a test of an hypothesis is conducted.

   In an initial experiment to assess the merits of using a newly developed filling material in bed pillows, 10 persons randomly selected from a group of volunteers agreed to test the new filling by actually using both the standard pillow and one made with the new filling in their homes. To avoid biases that might influence the volunteer's decision, both pillows were covered with the same material. One pillow carried the number "one" and the other the number "two." Only the experimenter knew which number represented the standard and which represented the new material. After one week's use, each volunteer stated his preference for one of the two pillows. If there is no underlying difference between the new and standard pillow, then the probability that a volunteer would prefer the new pillow to the standard would be $p = \frac{1}{2}$. If, on the other hand, this is not true and the new pillow has more desirable properties than the standard, then p, the probability that a volunteer prefers the new pillow, would be greater than $\frac{1}{2}$.

   Assuming that there is no difference between the pillows ($p = \frac{1}{2}$), it would be extremely unlikely that in a sample of $n = 10$ trials we would observe 9 or more people preferring the new pillow. From the table of binomial probabilities this probability, $p(9) + p(10)$, is .011. Therefore, more than 8 preferences for the new pillow would be sufficient evidence to reject the value $p = \frac{1}{2}$. Since y, the number of preferences, is a binomial random variable with possible values $0, 1, 2, \ldots, 10$, the possible outcomes can be divided into those (9 and 10) for which we agree to _____ the null hypothesis, and those $(0, 1, 2, \ldots, 8)$ for which we _____ the null hypothesis:

Margin answers:

p

probable
improbable

evidence

reject
accept

Possible values for y

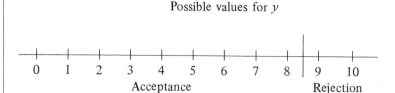

0 1 2 3 4 5 6 7 8 9 10

Acceptance   Rejection
Region   Region

The decision to reject the null hypothesis will be made if the observed value of y lies in the rejection region.

A statistical test of a theory possesses four elements:

1. There must be a theory to be tested, which we call the _____ _____, H_0. In our problem, H_0 declares that $p = \frac{1}{2}$. The objective of the test is to give the facts (data) a chance to refute H_0.

2. There must be an _____ hypothesis, H_a. If H_0 is false, then some alternative hypothesis is true. H_a generally expresses the experimenter's intuitive feeling about the true state of nature. In our example, if the new filling is better than the standard, then H_a would appropriately be $H_a: p > \frac{1}{2}$.

3. *The Test Statistic.* What information supplied by the data is relevant to a test of H_0? The statistic that reflects the true value of p is y, the number of _____ in 10 trials.

4. *The Rejection Region.* Certain values of the test statistic are more likely if H_a is true than if H_0 is true. A set of such values may be used as the _____ region for the test. Hence, if y takes on a value in the rejection region, we agree to (accept, reject) H_0. In our example, the expected number of preferences would be 5 if H_0 is true. However, if H_a is true and $p = .8$, then the expected number of preferences would be $np = 10(.8) = 8$. Thus, when $H_a: p > \frac{1}{2}$ is true, we would expect to obtain larger values of y, and the rejection region $y = 9$, 10 would be appropriate for our alternative.

The rejection region given as $y = 9$, 10 is not the only possible choice available to the experimenter. Someone requiring stronger evidence before rejecting H_0 as false might prefer the following assignment:

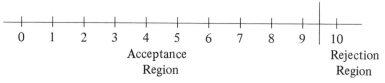

0 1 2 3 4 5 6 7 8 9 10

Acceptance   Rejection
Region   Region

Another person might be particularly interested in protecting himself against accepting H_0 when in fact p is greater than $\frac{1}{2}$, and could argue for the following assignment:

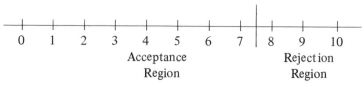

0 1 2 3 4 5 6 7 8 9 10

Acceptance   Rejection
Region   Region

null
hypothesis

alternative

preferences

rejection
reject

A sound choice among various reasonable rejection regions can be made after considering the possible errors that can be made in a test of an hypothesis.

The following is called a decision table and looks at the two possible states of nature (H_0 and H_a) and the two possible decisions in a test of an hypothesis. Fill in the missing entries as either "correct" or "error":

Null Hypothesis	Decision	
	Reject H_0	Accept H_0
True	_____	correct
False	correct	_____

(margin: error / error)

An error of Type I is made when one rejects H_0 when H_0 is (true, false). An error of Type II is made when one fails to reject H_0 when H_a is (true, false). In considering a statistical test of an hypothesis, it is essential to know the probabilities of committing errors of Type I and Type II when the test is used. We define

(margin: true / true)

$$\alpha = P \text{ [Type I error]} = P \text{ [reject } H_0 \text{ when } H_0 \text{ true]}$$

$$\beta = P \text{ [Type II error]} = P \text{ [accept } H_0 \text{ when } H_a \text{ true]}$$

For our example, let us look at α and β for the three rejection regions discussed above, using the tables of binomial probabilities from your text.
1. For the rejection region given as $y = 9, 10$,

$$\alpha = P \text{ [reject } H_0 \text{ when } H_0 \text{ true]}$$

$$= P \text{ [}y = 9 \text{ or } 10 \text{ when } p = \tfrac{1}{2}]$$

$$= 1 - P \text{ [}y \leqslant 8 \text{ when } p = \tfrac{1}{2}]$$

$$= 1 - \underline{\hspace{2cm}}$$

(margin: .989)

$$= \underline{\hspace{2cm}}.$$

(margin: .011)

If H_0 is false and if $p = .8$, then

$$\beta = P \text{ [accept } H_0 \text{ when } H_a \text{ true]}$$

$$= P \text{ [}y \leqslant 8 \text{ when } p = .8]$$

$$= \underline{\hspace{2cm}}.$$

(margin: .624)

2. For the second rejection region, $y = 10$,

$$\alpha = P \text{ [}y = 10 \text{ when } p = \tfrac{1}{2}]$$

$$= 1 - P \text{ [}y \leqslant 9 \text{ when } p = \tfrac{1}{2}]$$

.999

$$= 1 - \underline{\hspace{2cm}}$$

.001

$$= \underline{\hspace{2cm}},$$

while if H_0 is false and if $p = .8$,

$$\beta = P\,[y \leqslant 9 \text{ when } p = .8]$$

.893

$$= \underline{\hspace{2cm}}.$$

3. For the third rejection region, $y = 8, 9, 10$,

$$\alpha = P\,[y \geqslant 8 \text{ when } p = \tfrac{1}{2}]$$

$$= 1 - P\,[y \leqslant 7 \text{ when } p = \tfrac{1}{2}]$$

.945

$$= 1 - \underline{\hspace{2cm}}$$

.055

$$= \underline{\hspace{2cm}}.$$

If H_0 is false and if $p = .8$, then

$$\beta = P\,[y \leqslant 7 \text{ when } p = .8]$$

.322

$$= \underline{\hspace{2cm}}.$$

decreases

.05

Notice that both α and β depend upon the rejection region employed and that when the sample size n is fixed, α and β are inversely related: as one increases, the other _____. Increasing the sample size provides more information on which to make the decision and will reduce the Type II error probability. Since these two quantities measure the risk of making an incorrect decision, the experimenter chooses reasonable values for α and β and then chooses the rejection region and sample size accordingly. Since experimenters have found that a 1-in-20 chance of a Type I error is usually tolerable, common practice is to choose $\alpha \leqslant$ _____ and a sample size n large enough to provide the desired control of Type II error.

Example 6.17
Twenty students were tested for reaction time before and after lunch. Seventeen of the students showed increased reaction time after lunch. Is this sufficient evidence to indicate that reaction times are increased after lunch?

½
> ½

Solution
We begin by putting this problem into the context of a test of an hypothesis concerning p, the probability that reaction time has increased after lunch. The twenty students will be considered as $n = 20$ trials in a binomial experiment with $y = 17$. If eating lunch does not affect reaction time, then p = _____; but if eating lunch causes reaction time to increase, then p _____.

1. The *hypotheses* to be tested are

$H_0: p = \frac{1}{2}$ versus

$H_a: p > \frac{1}{2}$.

2. The *test statistic* will be y, the number of students exhibiting increased reaction time after lunch.

3. To choose a *rejection region*, we note that if H_0 is true, $\mu = np = 20(\frac{1}{2})$ = _____, while if H_a is true, and say $p = .7$, then $\mu = np = 20(.7)$ = _____. If $p = .9$, $\mu = np = 20(.9)$ = _____. If H_a is true, we should expect to obtain _____ values of y. Using your table of binomial probabilities with $n = 20$ and $p = \frac{1}{2}$, we need to find a cutoff number, a, such that

$\alpha = P [y \geqslant a \text{ when } p = \frac{1}{2}] \leqslant .05.$

Complete the entries below:

a	$P [y \geqslant a \text{ when } p = \frac{1}{2}]$
20	.000
19	.000
18	.000
17	.001
16	_____
15	_____
14	_____
13	_____

The largest rejection region with $\alpha \leqslant .05$ would consist of the values $y =$ _____, _____, ..., 20. (Note that an experimenter might be willing to include $y = 14$ in the rejection region and use $\alpha = .058$.)

4. Since the observed value of $y = 17$ lies in the rejection region, we reject the _____ _____ and conclude that eating lunch (does, does not) significantly increase reaction time.

5. It is worthwhile to note that both types of errors cannot be made at the same time. If we decide to reject H_0, the only error applicable is the _____ error. If we decide to accept H_0, the only possible error is _____. For this problem, we could then put a measure of goodness on our inference by noting that we conclude that eating lunch increases reaction time with probability _____ of being incorrect.

Right column answers:

10
14; 18
larger

.006
.021
.058
.132

15; 16

null hypothesis
does

Type I
Type II

.021

Self-Correcting Exercises 6D

1. In an experiment to determine the influence of suggestion on a subject's response, under a controlled situation 20 objects were shown to each of two subjects and the subjects each estimated the distance to the object aloud. Unknown to subject B, subject A (who always responded first) was part of the experiment and was told to overestimate the distance for the first ten objects and underestimate the distance for the last ten objects. The experimenter recorded the number of times that subject B agreed with subject A in either overestimating or underestimating the true

distance; he found that subject B agreed with subject A 16 out of 20 times. Is this sufficient evidence to conclude that there is an influence due to suggestion?

2. The supervisor of an elementary school is concerned because in the past tests revealed that only 80% of the first graders have grasped the concept of "size." A teaching intern devises a new method for teaching the concept and uses it on a class of 25 first graders. The supervisor administers a test and finds that 22 first graders understand the concept of size. Does this provide sufficient evidence to conclude that the new method is more successful in teaching the concept of size? Use $\alpha \leqslant .05$.

3. While ordering a new shipment of shirts, the owner of a men's shop was told that the demand for a new color was anticipated to comprise about 40% of sales during the next season. The owner ordered his shipment in line with this 40% figure. If a random inspection of 15 sales slips involving the sale of a shirt revealed that four of these sales involved shirts of the new color, could the owner conclude that the 40% figure was actually too high?

EXERCISES

1. A subject is taught to do a task in two different ways. Studies have shown that, when subjected to mental strain and asked to perform the task, the subject most often reverts to the method first learned, regardless of whether it was easier or more difficult than the second. If the probability that a subject returns to the first method learned is .8 and six subjects are tested, what is the probability that at least 5 of the subjects revert to their first learned method when asked to perform their task under mental strain?

2. The taste test for PTC (phenylthiourea) is a favorite exercise for every human genetics class. It has been established that a single gene determines the characteristic, and 70% of the American population are "tasters," while 30% are "non-tasters." If 20 people are randomly chosen and administered the test,
    a. Give the probability distribution of y, the number of "non-tasters" out of the 20 chosen.
    b. Using appropriate tables, find $P[y \leqslant 7]$.
    c. Find $P[3 < y \leqslant 8]$.

3. A multiple-choice test offers four alternative answers to each of 100 questions. In every case there is but one correct answer. Bill responded correctly to each of the first 76 questions when he noted that just 20 seconds remained in the test period. He quickly checked an answer at random for each of the remaining 24 questions without reading them.
    a. What is Bill's expected number of correct answers?
    b. If the instructor assigns a grade by taking 1/3 of the wrong from the number marked correctly, what is Bill's expected grade?

4. Shipments of refrigerators are accepted in accordance with the sampling plan $n = 2, a = 0$.
    a. Find the probability of accepting a lot with fraction defective, $p = .01$.
    b. Find the probability of accepting a lot with fraction defective, $p = .20$.

5. Refer to Exercise 4.
    a. Find a sampling plan with sample size $n = 20$ which has approximately

the same probability of accepting a lot with fraction defective, $p = .01$, as the plan $n = 2, a = 0$.

b. What is the probability under the plan determined in part a. of accepting a lot which has fraction defective, $p = .20$?

c. If you were purchasing refrigerators by the lot, what advantage would there be in using the plan determined in part a. rather than the plan $n = 2, a = 0$? What disadvantages can you cite for using the plan with sample size $n = 20$?

6. A coroner's null hypothesis is H_0: "This man is alive." If you were "this man" would you prefer a test with $\alpha = .05$ and $\beta = .001$ or a test with $\alpha = .001$ and $\beta = .05$? Explain your preference.

7. A new method of packaging brand A candy has been proposed as a means of increasing sales. It is known that approximately 40% of the potential customers now purchase brand A. If at least six of the next ten customers (each of whom is given a choice of brand A in the new package or one of its competitors) select brand A, we shall conclude that the new packaging method is effective in increasing sales.

a. State H_0 in terms of p, the probability that a given customer will select brand A.

b. Find α for this experiment. (Use Table 1, text.)

c. State H_a in terms of p.

d. Find β for $H_a: p = .6$. (Use Table 1, text.)

8. It is thought that cottage cheese batches in two tanks (tank A and tank B) are equally desirable. Let p denote the probability that a given taster will express a preference for the cottage cheese in tank A. To test the null hypothesis, $H_0: p = 1/2$, against the alternative, $H_a: p \neq 1/2$, each member of a panel of ten tasters is asked to judge which cottage cheese is the more desirable. Let y denote the number of tasters who will state a preference for the product in tank A. Suppose that the rejection region consists of the values $y = 0, 1, 9$ and 10.

a. Describe the Type I error in terms of the cheeses.

b. Describe the Type II error in terms of the cheeses.

c. Find α for the above test.

d. Find β if indeed $p = .60$.

e. Find β if indeed $p = .90$.

f. Find β if indeed $p = .99$.

g. From the answers recorded for parts d, e, and f state whether β is larger when p is close to the value specified in H_0 or when p is grossly different from the value specified in H_0.

9. Sacks of grapefruit at a certain food store contain, according to the label, 20 pounds of fruit. The produce manager claims that only 10% of these sacks fail to contain at least 20 pounds of fruit. To test this claim, a sample of 25 sacks is selected at random from his stock, and four sacks fail to contain at least 20 pounds of fruit. Is this sufficient evidence to refute the manager's claim? Use $\alpha \leq .05$.

10. The probability that a patient will recover from a certain rare disease is known to be .20. To test the effectiveness of a new drug it is administered to each of 20 patients afflicted with this disease. If 9 of the 20 patients recover, does the drug appear to significantly increase the recovery rate? Use $\alpha \leq .05$.

Chapter   7

THE NORMAL PROBABILITY DISTRIBUTION

7.1 Introduction

Recall that a random variable was defined to be a numerically valued
_____ defined over a _____ _____. Random   function; sample space
variables can be divided into two categories, (1) _____ random   discrete
variables and (2) _____ random variables. We have confined our   continuous
discussion to problems concerning discrete random variables which can take a
_____ or countable _____ of values. In contrast, a   finite; infinity
_____ random variable can assume any of the values associated   continuous
with an interval on the real line.

  While the probability distribution or frequency distribution of a
_____ random variable can be represented by a relative frequency   discrete
histogram, the probability distribution for a _____ random variable   continuous
is represented by a smooth curve. The probability distribution for the normal
random variable y is

$$\left[\quad f(y) = \frac{1}{\sigma\sqrt{2\pi}}\, e^{-\frac{1}{2}\left(\frac{y-\mu}{\sigma}\right)^{2}} \quad -\infty < y < \infty \quad \right]$$

which represents the "_____-shaped" curve shown below.   bell

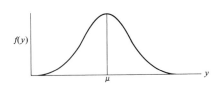

The symbols used in the function $f(y)$ are defined as follows:
1. π and e are irrational numbers whose approximate values are 3.1416 and
    2.7183, respectively.
2. μ and σ are constants which represent the population _____ and   mean
    _____ _____, respectively.   standard deviation
Encountering a random variable whose values can be extremely small (a large

negative value) or extremely large might at first be disconcerting to the student who has heard that heights, weights, response times, and errors of measurements are approximately normally distributed. Surely we do not have heights, weights, or times that are less than zero! Certainly not, but 99.7% of the distribution of a normally distributed random variable lies within the interval $\mu \pm 3\sigma$. In the case of heights or weights this interval almost always encompasses positive values. Keep in mind this curve is merely a *model* that approximates an actual distribution of measurements. Its great utility lies in the fact that it *can* be used effectively as a model for so many types of measurements.

7.2 The Central Limit Theorem

Consider an experiment whereby the distance traveled by a shell fired from a given cannon is measured and recorded. If none of the settings on the cannon were changed and this experiment were repeated a large number of times, a frequency histogram of these distances would probably exhibit the mound-shaped distribution characteristic of the normally distributed random variable. Why should this be the case? The cannon undoubtedly has some sort of average firing distance, but each measurement would deviate from this value due perhaps to errors in measurement by the person recording the distance, the air termperature or humidity at the time of firing, the exact amount of gun powder in the chamber, the angle of the cannon when fired, and so on. Hence, any one distance would consist of an average distance modified by the addition of random errors, which might be either positive or negative.

The Central Limit Theorem loosely stated says that sums or averages are approximately normally distributed with a mean and standard deviation that depend upon the sampled population. By thinking of the error in the distance measured (the deviation from the "average distance") as a *sum* of various effects in which small errors are highly likely and large errors are highly improbable, the Central Limit Theorem helps explain the apparent normality of the shell distances.

Further, and just as important, the Central Limit Theorem assures us that sample means will be approximately normally distributed with a mean and variance that depend upon the population from which the sample has been drawn. This aspect of the Central Limit Theorem will be the focal point for making inferences about populations based upon random samples when the sample size is large.

> *The Central Limit Theorem (1):* If random samples of n observations are drawn from a population with finite mean μ and standard deviation σ, then when n is large, the sample mean $\bar{y}$ will be approximately normally distributed with mean μ and standard deviation $\sigma/\sqrt{n}$.

The Central Limit Theorem could also be stated in terms of the sum of the measurements, Σy_i.

> *The Central Limit Theorem (2):* If random samples of n observa-

tions are drawn from a population with finite mean μ and standard deviation σ, then when n is large, $\Sigma\, y_i$ will be approximately normally distributed with mean $n\mu$ and standard deviation $\sigma\sqrt{n}$.

In both cases the approximation to normality becomes more and more accurate as n becomes large.

The Central Limit Theorem is important for two reasons.
1. It partially explains *why* certain measurements possess approximately a _____ distribution.   normal
2. Many of the _____ used in making inferences are sums or means   estimators
of sample measurements and thus possess approximately _____   normal
distributions for large samples. The student should notice that the Central Limit Theorem (does, does not) specify that the sample measurements   does not
come from a normal population. The population (could, could not) have a   could
frequency distribution that is flat or skewed or is non-normal in some other way. It is the *sample mean* that behaves as a random variable having an approximately normal distribution.

To clarify a point we note that the sample mean $\bar{y}$ computed from a random sample of n observations drawn from any population with mean μ and standard deviation σ always has a mean equal to μ and a standard deviation equal to $\sigma/\sqrt{n}$. This result is not due to the Central Limit Theorem. The important contribution of the theorem lies in the fact that when n, the sample size, is *large,* we may approximate the distribution of $\bar{y}$ with a *normal* probability distribution.

7.3 Random Samples

Since it is the sample that provides the information that is used in inference making, we must be duly careful about the selection of the elements in the sample so that we do not systematically exclude or include certain elements of the population in our sampling plan. The sample should be representative of the population being sampled.

We call a sample that has been drawn without bias a *random sample.* This is a shortened way of saying that the sample has been drawn in a random manner. A sample of size n is said to have been randomly drawn if each possible sample of size n has the same chance of being selected.

If a population consists of N elements and we wish to draw a sample of size n from this population, there are

$$C_n^N = \frac{N!}{n!\,(N-n)!}$$

samples to choose from. A random sample in this situation would be one drawn in such a manner that each sample of size n had the same chance of being drawn, namely, $(C_n^N)^{-1}$ or $1/C_n^N$.

Although perfect random sampling is difficult to achieve in practice, there are several methods available for selecting a sample that will satisfy the conditions of random sampling when N, the population size, is not too large.

1. *Method A.* List all the possible samples and assign them numbers. Place each of these numbers on a chip or piece of paper and place them in a bowl. Drawing one number from the bowl will select the random sample to be used.
2. *Method B.* Number each of the N members of the population. Write each of these numbers on a chip or slip of paper and place them in a bowl. Now draw n numbers from the bowl and use the members of the population having these numbers as elements to be included in the sample.
3. *Method C.* A useful technique for selecting random samples is one in which a table of random numbers is used to replace the chance device of drawing chips from a bowl.

Why is it so important that the sample be randomly drawn? From the practical point of view, one would want to keep the experimenter's biases out of the selection, and at the same time keep the sample as representative of the

population

_____ as possible. From the statistical point of view, we can assess the probability of observing a random sample and hence make valid

inferences

cannot

_____ about the parent population. If the sample is nonrandom, its probability (can, cannot) in general be determined and hence no valid inferences can be made from it.

7.4 The Standard Normal Probability Distribution

Probability is the vehicle through which we are able to make inferences about a population in the form of either estimation or decisions. To make inferences about a normal population, we must be able to compute or otherwise find the probabilities associated with a normal random variable. However, since the probability distribution for a normal random variable y depends upon the

$\mu; \sigma$

population parameters _____ and _____, one would be required to calculate anew the probabilities associated with y each time a new value for μ or σ was encountered. We resort to a standardization process whereby we

standard

convert a normal random variable y to a _____ normal random variable z which represents the distance of y from its mean, μ, in units of the standard deviation, σ. To standardize a normal random variable y we use the following procedure:

1. From y, subtract its mean, μ.

$$y - \mu$$

This results in the signed distance of y from its mean, a negative sign indicating that y is to the left of μ while a positive sign indicates y is to the right of μ.

2. Now divide by σ.

$$\frac{y - \mu}{\sigma}$$

Dividing by σ converts the signed distance from the mean to the number of standard deviations to the right or left of μ.

3. Define

$$z = \frac{y - \mu}{\sigma}.$$

z is the standard normal variable having the standardized normal distribution with mean 0 and standard deviation 1.
Given the curve representing the distribution of a continuous random variable
y, the probability that $a \leqslant y \leqslant b$ is represented by the _____
under the curve between the points _____ and _____. Hence in
finding probabilities associated with a standardized normal variable z, we
could refer directly to the areas under the curve. These areas are tabulated for
your convenience in Table 3 of your text.
    Since the standardized normal distribution is _____ about the
mean zero, half of the area lies to the left of zero and half to the right of zero,
and the areas to the left of the mean $z = 0$ can be calculated by using the corresponding and equal area to the right of $z = 0$. Hence Table 3 exhibits areas
only for positive values of z correct to the nearest hundredth. Table 3 gives
the area between $z = 0$ and a specified value of z, say z_0. A convenient notation
used to designate the area between $z = 0$ and z_0 is $A(z_0)$.

area
$a; b$

symmetric

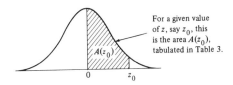

For a given value
of z, say z_0, this
is the area $A(z_0)$,
tabulated in Table 3.

a. For $z = 1$, the area between $z = 0$ and $z = 1$ is $A(z = 1) = A(1) = .3413$.
b. For $z = 2$, $A(z = 2) = A(2) =$ _____.
c. For $z = 1.6$, $A(1.6) =$ _____.
d. For $z = 2.4$, $A(2.4) =$ _____.
Now try reading the table for values of z given to two decimal places.
e. For $z = 2.58$, $A(2.58) =$ _____.
f. For $z = 0.75$, $A(0.75) =$ _____.
g. For $z = 1.69$, $A(1.69) =$ _____.
h. For $z = 2.87$, $A(2.87) =$ _____.
We will now find probabilities associated with the standard normal random
variable z by using Table 3.

.4772
.4452
.4918

.4951
.2734
.4545
.4979

Example 7.1
Find the probability that z is greater than 1.86, that is, $P[z > 1.86]$.

Solution
Illustrate the problem with a diagram as follows:

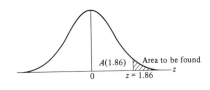

$A(1.86)$ Area to be found

0 $z = 1.86$

1. The total area to the right of $z = 0$ is equal to .5000.

.4686
.5000

2. From Table 3, $A(1.86) = $ _____.
3. Therefore, the shaded area is found by subtracting $A(1.86)$ from _____.
4. Hence,

$$P[z > 1.86] = .5000 - A(1.86)$$

.4686

$$= .5000 - \underline{\hspace{2cm}}$$

.0314

$$= \underline{\hspace{2cm}}.$$

Example 7.2
Find $P[z < -2.22]$.

Solution
Illustrate the problem with a diagram.

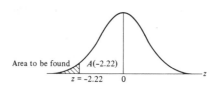

Area to be found $A(-2.22)$
$z = -2.22$ 0

1. Using the symmetry of the normal distribution, $A(-2.22) = A(2.22)$

.4868
left

$= $ _____. The minus value of z indicates that you are to the (left, right) of the mean, $z = 0$.

.4868

2. $P[z < -2.22] = .5000 - \underline{\hspace{2cm}}$

.0132

$$= \underline{\hspace{2cm}}.$$

Example 7.3
Find $P[-1.21 < z < 2.43]$.

Solution
Illustrate the problem with a diagram.

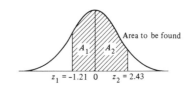

Area to be found
A_1 A_2
$z_1 = -1.21$ 0 $z_2 = 2.43$

$$P[-1.21 < z < 2.43] = P[-1.21 < z < 0] + P[0 < z < 2.43]$$

.3869; .4925

$$= \underline{\hspace{2cm}} + \underline{\hspace{2cm}}$$

.8794

$$= \underline{\hspace{2cm}}.$$

A second type of problem that arises is that of finding a value of z, say z_0, such that a probability statement about z will be true. We explore this type of problem with examples.

Example 7.4
Find the value of z_0 such that

$$P[0 < z < z_0] = .3925.$$

Solution
Once again, illustrate the problem with a diagram and list pertinent information.

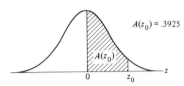

$A(z_0) = .3925$

$A(z_0)$

0   z_0

1. Search Table 3 until the area .3925 is found. The value such that

     $A(z_0) = .3925$ is $z_0 = $ _____   1.24

2. $P[0 < z < $ _____ $] = .3925.$ 1.24

Example 7.5
Find the value of z_0 such that

$$P[z > z_0] = .2643.$$

Solution
Illustrate the problem and list pertinent information.

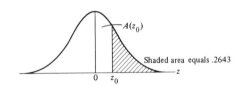

$A(z_0)$

Shaded area equals .2643

0   z_0

1. $A(z_0) = .5000 - .2643 = $ _____ .   .2357

2. The value of z_0 such that

        $A(z_0) = $ _____ is $z_0 = $ _____ .   .2357; 0.63

3. $P[z > $ _____ $] = .2643.$ 0.63

Self-Correcting Exercises 7A

1. Find the following probabilities associated with the standard normal random variable z:

a. $P[z > 2.1]$.

b. $P[z < -1.2]$.

c. $P[.5 < z < 1.5]$.

d. $P[-2.75 < z < -1.70]$.

e. $P[-1.96 < z < 1.96]$.

f. $P[z > 1.645]$.

2. Find the value of z, say z_0, such that the following probability statements are true:

a. $P[z > z_0] = .10$.

b. $P[z < z_0] = .01$.

c. $P[-z_0 < z < z_0] = .95$.

d. $P[-z_0 < z < z_0] = .99$.

3. An auditor has reviewed the financial records of a hardware store and has found that their billing errors follow a normal distribution with mean and standard deviation equal to $0 and $1 respectively.
   a. What proportion of the store's billings are in error by more than $1?
   b. What is the probability that a billing represents an overcharge of at least $1.50?
   c. What is the probability that a customer has been undercharged from $0.50 to $1.00?
   d. Within what range would 95% of the billing errors lie?
   e. Of the extreme undercharges, 5% would be at least what amount?

7.5 Use of the Table for the General Normal Variable y

We can now proceed to find probabilities associated with any normal random variable y having mean μ and standard deviation σ. This is accomplished by converting the random variable y to the standard normal random variable z, and then working the problem in terms of z.

Since probability statements are written in the form of inequalities, the reader is reminded of two facts. A statement of inequality is maintained if
1. the same number is subtracted from each member of the inequality and/or
2. each member of the inequality is divided by the same *positive* number.

Example 7.6
The following are equivalent statements about y:

1. $70 < y < 95$.

2. $\quad (70 - 15) < (y - 15) < (95 - 15).$

3. $\quad \dfrac{70 - 15}{5} < \dfrac{y - 15}{5} < \dfrac{95 - 15}{5}.$

4. $\quad 11 < \dfrac{y - 15}{5} < 16.$

Example 7.7
Let y be a normal random variable with mean $\mu = 100$ and standard deviation $\sigma = 4$. Find $P[92 < y < 104]$.

Solution
Recalling that $z = (y - \mu)/\sigma$, we can apply rules 1 and 2 to convert the probability statement about y to one about the standard normal random variable z.

1. $\quad P[92 < y < 104] = P[92 - 100 < y - 100 < 104 - 100]$

$$= P\left[\dfrac{92 - 100}{4} < \dfrac{y - 100}{4} < \dfrac{104 - 100}{4}\right]$$

$$= P[-2 < z < 1].$$

2. The problem now stated in terms of z is readily solved using the methods of section 7.4.

$$P[92 < y < 104] = P[-2 < z < 1]$$

$$= A(-2) + A(1)$$

$$= \underline{\hspace{2cm}} + \underline{\hspace{2cm}} \qquad \qquad .4772; .3413$$

$$= \underline{\hspace{2cm}}. \qquad \qquad .8185$$

Example 7.8
Let y be a normal random variable with mean 100 and standard deviation 4. Find

$$P[93.5 < y < 105.2].$$

Solution

$$P[93.5 < y < 105.2] = P\left[\dfrac{93.5 - 100}{4} < z < \dfrac{105.2 - 100}{4}\right]$$

$$= P[-1.63 < z < \underline{\hspace{2cm}}] \qquad \qquad 1.30$$

$$= \underline{\hspace{2cm}} + \underline{\hspace{2cm}} \qquad \qquad .4484; .4032$$

$$= \underline{\hspace{2cm}}. \qquad \qquad .8516$$

Self-Correcting Exercises 7B

1. If y is normally distributed with mean 10 and variance 2.25, evaluate the following probabilities:

    a. $P[y > 8.5]$.

    b. $P[y < 12]$.

    c. $P[9.25 < y < 11.25]$.

    d. $P[7.5 < y < 9.2]$.

    e. $P[12.25 < y < 13.25]$.

2. An industrial engineer has found that the standard household light bulbs produced by a certain manufacturer have a useful life which is normally distributed with a mean of 250 hours and a variance of 2500. What is the probability that a randomly selected bulb from this production process will have a useful life
   a. in excess of 300 hours?
   b. between 190 and 270 hours?
   c. not exceeding 260 hours?
   d. Ninety percent of the bulbs have a useful life in excess of how many hours?
   e. The probability is .95 that a bulb does not have a useful life in excess of how many hours?

3. Scores on a trade school entrance examination form exhibit the characteristics of a normal distribution with mean and standard deviation of 50 and 5, respectively. What proportion of the scores on this examination form would be
   a. greater than 60?
   b. less than 45?
   c. between 35 and 65?
   d. If to be considered eligible for a place in the incoming class an applicant must score beyond the 75th percentile on this form, what score must an applicant have to be eligible?

7.6 The Normal Approximation to the Binomial Distribution

For large values of n, the binomial probabilities, $p(y) = C_y^n p^y q^{n-y}$ are very tedious to compute. Is there an alternative to long computations? There is, as we shall see.

For a binomial experiment consisting of n trials, let

$$y_1 = 1 \quad \text{if trial one is a success}$$
$$0 \quad \text{if trial one is a failure}$$

$y_2 = 1$ if trial two is a success

     0   if trial two is a failure

.
.
.

$y_n = 1$ if trial n is a success

     0   if trial n is a failure.

Then y, the number of successes in n trials, can be thought of as a sum of n independent random variables each with mean equal to _____ and a variance equal to _____. For p pq

$$y = y_1 + y_2 + \ldots + y_n$$

the _____ _____ Theorem says this sum is _____ normally distributed with mean _____ and variance _____. Hence when n is _____, we can approximate the distribution of a bi- nomial random variable with the distribution of a normal random variable whose mean and variance are identical to those for the binomial random variable. When can we reasonably apply the normal approximation? For small values of n and values of p close to zero or one, the binomial distribution will exhibit a "pile-up" around $y =$ _____ or $y =$ _____. The data will not be _____-shaped and the normal approximation will be poor. For a normal random variable, _____% of the measurements will be within the interval $\mu \pm 2\sigma$. For $\mu = np$ and $\sigma = \sqrt{npq}$, the interval $np \pm 2\sqrt{npq}$ should be within the bounds of the binomial random variable y, or within the interval $(0, n)$ to obtain reasonably good approximations to the binomial probabilities.

Central Limit; approximately
np; npq
large

0; n
bell
95

To show how the normal approximation is used, let us consider a binomial random variable y with $n = 8$ and $p = 1/2$ and attempt to approximate some binomial probabilities with a normal random variable having the same mean, $\mu = np$, and variance, $\sigma^2 = npq$, as the binomial y. In this case

$$\mu = np = 8(1/2) = \underline{\hspace{1.5cm}}.$$

4

$$\sigma^2 = npq = 8(1/2)(1/2) = \underline{\hspace{1.5cm}}.$$

2

Note that the interval

$$\mu \pm 2\sigma = 4 \pm 2\sqrt{2} = (1.2, 6.8)$$

is contained within the interval $(0, 8)$; therefore our approximations should be adequate. Consider the following diagrammatic representation of the approximation where $p(y)$ is the frequency distribution for the binomial random variable and $f(y)$ is the frequency distribution for the corresponding normal random variable y.

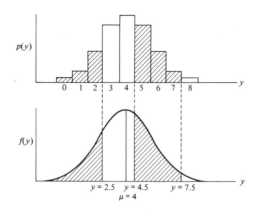

Example 7.9
Find $P[y < 3]$ using the normal approximation.

Solution
$P[y < 3]$ for the binomial random variable with mean $\mu = 4$ and $\sigma = \sqrt{2}$ corresponds to the shaded bars in the histogram over $y = 0, 1,$ and 2. The approximating probability corresponds to the shaded area in the normal distribution with mean 4 and standard deviation $\sqrt{2}$ to the left of $y = 2.5$.
We proceed as follows:

$$P[y < 3] \approx P[y < 2.5]$$

$$= P\left[\frac{y - 4}{\sqrt{2}} < \frac{2.5 - 4}{\sqrt{2}}\right]$$

$$= P[z < -1.06]$$

$$= .5000 - A(-1.06)$$

.3554

$$= .5000 - \underline{\hspace{2cm}}$$

.1446

$$= \underline{\hspace{2cm}}.$$

Example 7.10
Find $P[5 \leqslant y \leqslant 7]$.

Solution
For the binomial random variable with mean 4 and standard deviation $\sqrt{2}$,

5

$P[5 \leqslant y \leqslant 7]$ corresponds to the shaded bars over $y = \underline{\hspace{1.5cm}}$,

6; 7

$\underline{\hspace{1.5cm}}$, and $\underline{\hspace{1.5cm}}$. This corresponds in turn to the shaded area for the approximating normal distribution with mean 4 and standard deviation

4.5; 7.5

$\sqrt{2}$ between $y = \underline{\hspace{1.5cm}}$ and $y = \underline{\hspace{1.5cm}}$. Therefore,

$$P[5 \leqslant y \leqslant 7] \approx P[4.5 < y < 7.5]$$

$$= P\left[\frac{4.5 - 4}{\sqrt{2}} < \frac{y - 4}{\sqrt{2}} < \frac{7.5 - 4}{\sqrt{2}}\right]$$

$$= P[.35 < z < 2.47]$$

$$= A(2.47) - A(.35)$$

$$= \underline{\hspace{2cm}} - \underline{\hspace{2cm}}$$ .4932; .1368

$$= \underline{\hspace{2cm}} .$$ .3564

Notice that we used $P[y < 2.5]$ to approximate the binomial probability $P[y < 3]$. In like manner we used $P[4.5 < y < 7.5]$ to approximate the binomial probability $P[5 \leqslant y \leqslant 7]$. The addition or subtraction of 0.5 is called *correction for continuity* since we are approximating a discrete probability distribution with a probability distribution that is continuous.

A student may become confused as to whether 0.5 should be added or subtracted in the process of approximating binomial probabilities. A common sense rule that always works is to examine the binomial probability statement carefully, and determine which values of the binomial random variable are included in the statement. (Draw a picture if necessary.) The probabilities associated with these values correspond to the bars in the histogram centered over them. Locating the end points of the bars to be included determines the values needed for the approximating normal random variable.

Example 7.11
Suppose y is a binomial random variable with $n = 400$ and $p = 0.1$. Use the normal approximation to binomial probabilities to find:

a. $\qquad P[y > 45]$.

b. $\qquad P[y \leqslant 32]$.

c. $\qquad P[34 \leqslant y \leqslant 46]$.

Solution
If y is binomial, then its mean and variance are

$$\mu = np = 400(.1) = \underline{\hspace{2cm}} .$$ 40

$$\sigma^2 = npq = 400(.1)(.9) = \underline{\hspace{2cm}} .$$ 36

a. To find $P[y > 45]$, we need the probabilities associated with the values $46, 47, 48, \ldots, 400$. This corresponds to the bars in the binomial histogram beginning at $\underline{\hspace{2cm}}$. Hence 45.5

$$P[y > 45] \approx P[y > 45.5]$$

$$= P\left[z > \frac{45.5 - 40}{6}\right]$$

$$= P[z > .92]$$

.1788

$$= \underline{\hspace{2cm}}.$$

b. To find $P[y \leqslant 32]$, we need the probabilities associated with the values $0, 1, 2, \ldots$, up to and including $y = 32$. This corresponds to finding the area in the binomial histogram to the left of _____ . Hence

32.5

$$P[y \leqslant 32] \approx P[y < 32.5]$$

$$= P\left[z < \frac{32.5 - 40}{6}\right]$$

-1.25

$$= P[z < \underline{\hspace{1.5cm}}]$$

.1056

$$= \underline{\hspace{2cm}}.$$

c. To find $P[34 \leqslant y \leqslant 46]$, we need the probabilities associated with the values beginning at 34 up to and including 46. This corresponds to finding the area in the histogram between _____ and _____ . Hence

33.5; 46.5

$$P[34 \leqslant y \leqslant 46] \approx P[33.5 < y < 46.5]$$

$$= P\left[\frac{33.5 - 40}{6} < z < \frac{46.5 - 40}{6}\right]$$

-1.08; 1.08

$$= P[\underline{\hspace{1.5cm}} < z \underline{\hspace{1.5cm}}]$$

.7198

$$= \underline{\hspace{2cm}}.$$

Notice that the interval $\mu \pm 2\sigma$ or 40 ± 12 is well within the binomial range of 0 to 400, so that these approximate probabilities should be reasonably accurate.

Self-Correcting Exercises 7C

1. For a binomial experiment with $n = 20$ and $p = .7$, calculate $P[10 \leqslant y \leqslant 16]$
   a. using the binomial tables.
   b. using the normal approximations.
2. If the median income in a certain area is claimed to be $12,000, what is the probability that 37 or fewer of 100 randomly chosen wage-earners from this area have incomes less than $12,000? Would the $12,000 figure seem reasonable if your sample actually contained 37 wage-earners whose income was less than $12,000?
3. A large number of seeds from a certain species of flower are collected and mixed together in the following proportions according to the color of the flowers they will produce: 2 red, 2 white, 1 blue. If these seeds are mixed and then randomly packaged in bags containing about 100 seeds, what is the probability that a bag will contain

a. at most 50 "white" seeds?

b. at least 65 seeds that are not "white"?

c. at least 25 but at most 45 "white" seeds?

4. Refer to Exercise 3. Within what limits would you expect the number of white seeds to lie with probability .95?

EXERCISES

1. Find the following probabilities for the standard normal variable z:

a. $P[z < 1.9]$.

b. $P[1.21 < z < 2.25]$.

c. $P[z > -0.6]$.

d. $P[-2.8 < z < 1.93]$.

e. $P[-1.3 < z < 2.3]$.

f. $P[-1.62 < z < 0.37]$.

2. Find the value of z, say z_0, such that the following probability statements are true:

a. $P[z > z_0] = .2420$.

b. $P[-z_0 < z < z_0] = .9668$.

c. $P[-z_0 < z < z_0] = .90$.

d. $P[z < z_0] = .9394$.

3. If y is distributed normally with mean 25 and standard deviation 4, find

a. $P[y > 21]$.

b. $P[y < 30]$.

c. $P[15 < y < 35]$.

d. $P[y < 18]$.

4. A sidewalk interviewer stopped three men who were walking together, asked their opinions on some topical subjects, and found their answers quite similar. Would you consider the interviewer's selection to be random in this case? Is it surprising that similar answers were given by these three men?

5. A pre-election poll taken in a given city indicated that 40% of the voting

public favored candidate A, 40% favored candidate B, and 20% were as yet undecided. If these percentages are true, in a random sample of 100 voters what is the probability that

a. at most 50 voters in the sample prefer candidate A?

b. at least 65 voters in the sample prefer candidate B?

c. at least 25 but at most 45 voters in the sample prefer candidate B?

6. On a well-known college campus, the student automobile registration revealed that the ratio of small to large cars (as measured by engine displacement) is 2 to 1. If 72 car owners are chosen at random from the student body, find the probability that this group includes at most 46 owners of small cars.

7. A psychological "introvert-extrovert" test produced scores which had a normal distribution with mean and standard deviation 75 and 12, respectively. If we wish to designate the *highest* 15% as extrovert, what would be the proper score to choose as the cut-off point?

8. A manufacturer's process for producing steel rods can be regulated so as to produce rods with an average length μ. If these lengths are normally distributed with a standard deviation of 0.2 inches, what should be the setting for μ if one wants at most 5% of the steel rods to have a length greater than 10.4 inches?

9. For a given type of cannon and a fixed range setting, the distance that a shell fired from this cannon will travel is normally distributed with a mean and standard deviation of 1.5 and 0.1 miles, respectively. What is the probability that a shell will travel

a. farther than 1.72 miles?

b. less than 1.35 miles?

c. at least 1.45 miles but at most 1.62 miles?

d. If three shells are fired, what is the probability that all three will travel farther than 1.72 miles?

10. In introducing a new breakfast sausage to the marketing public, an advertising campaign claimed that 7 out of 10 shoppers would prefer these new sausages over other brands. If 100 people were randomly chosen, and the advertiser's claim is true, what is the probability that

a. at most 65 people preferred the new sausages?

b. at least 80 people preferred the new sausages?

c. If only 60 people stated a preference for the new sausages, would this be sufficient evidence to indicate that the advertising claim is false and that, in fact, less than 7 out of 10 people would prefer the new sausages?

Chapter 8

STATISTICAL INFERENCE

8.1 Introduction

The objective of statistics is to make _____ about a _____ — *inferences; population*
based on information contained in a _____. Since populations are — *sample*
described by numerical descriptive measures, called _____ of the — *parameters*
population, one can make inferences about the population by making
inferences about its parameters. For example, the test of the effectiveness of
a new vaccine and lot acceptance sampling were inferences which resulted in
decisions concerning the binomial parameter p. In this chapter we consider
two methods for making inferences concerning population parameters:
estimation and decision making. The quantities to be used in making inferences will be sums or averages of the measurements in a random sample and
consequently will possess frequency distributions in repeated sampling that
are approximately _____ due to the _____ _____ — *normal; Central Limit*
Theorem.

One of the most important concepts to grasp is that inference making is a
two-step procedure. These steps are:
1. Making the inference.
2. Measuring its goodness.
A measure of the goodness of an inference is essential to enable the person
using the inference to measure its reliability. For example, we would wonder
whether the drug employed in a medical experiment was truly effective or
whether the drug was ineffective and favorable experimental response
occurred due to chance.

This chapter will be concerned with making inferences about four parameters:
1. μ, the mean of a population of continuous measurements.
2. $\mu_1 - \mu_2$, the difference between the means for two populations of continuous measurements.
3. p, the parameter of a dichotomous or binomial population.
4. $p_1 - p_2$, the difference in the parameters for two binomial populations.
We shall make inferences about each of the four parameters just mentioned in
one of three ways:

1. by estimating the value of the parameter with a point estimate and giving a bound on the error of estimation,
2. by estimating the value of the parameter with a confidence interval, or
3. by testing an hypothesis about the value of the parameter.

The techniques developed in the making of inferences about these four parameters will also be used to determine how large the sample must be to achieve the accuracy required by the experimenter.

Rather than follow the section numbers exactly as they appear in your text, we have grouped certain topics together and will consider the following more general sections:

1. Point estimation.
2. Interval estimation.
3. Choosing the sample size.
4. A large sample test of an hypothesis.

8.2 Estimation

estimation; estimator

Using the measurements in a sample to predict the value of one or more parameters of a population is called _____. An _____ is a rule that tells one how to calculate an estimate of a parameter based on the information contained in a sample. One can give many different estimators for a given population parameter. An estimator is most often expressed in terms of a mathematical formula that gives the estimate as a function of the sample measurements. For example, $\bar{y}$ is an *estimator* of the population parameter, μ. If a sample of $n = 20$ pieces of aluminum cable is tested for strength and the mean of the sample is $\bar{y} = 100.7$, then 100.7 is an *estimate* of the population mean strength, μ. The estimator of a parameter is usually designated by placing a "hat" over the parameter to be estimated. Thus an estimator of μ would be $\hat{\mu} = \bar{y}$.

Estimates of a population parameter can be made in two ways:

1. The measurements in the sample can be employed to calculate a single number that is the estimate of the population parameter.
2. The measurements in the sample can also be used to calculate two points from which we acquire an estimate in the form of upper and lower limits within which the true value of the parameter is expected to lie. This type

interval

of estimate is called an _____ estimate since it defines an interval on the real line.

The goodness of an estimator is evaluated by observing its behavior in repeated sampling. Let us talk in general about some population parameter which we will denote as θ. An estimator $\hat{\theta}$ for the parameter θ will generate estimates in repeated sampling from the population and will produce a distribution of estimates (numerical values computed from these samples). This estimator would be considered good if the estimates cluster closely about θ.

unbiased

If the mean of the estimates is θ, then $\hat{\theta}$ is said to be an _____ estimator for θ and $E(\hat{\theta}) = \theta$. If the spread (variance) of $\hat{\theta}$ is smaller than that

minimum

of any other estimator, then $\hat{\theta}$ is said to have _____ variance.

Therefore, a *good estimator* should have the following properties:

unbiasedness

1. _____ .

minimum variance

2. _____ _____ .

The distributions obtained in repeated sampling are shown below for four different estimators of θ. Which estimator appears to possess the most desirable properties? _____

(d) $\hat{\theta}_4$

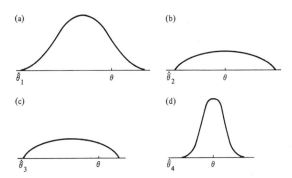

(a)

(b)

(c)

(d)

The properties of interval estimators are also determined by repeated sampling. Repeated use of an interval estimator generates a large number of _____ _____ for estimating θ. If an interval estimator were satisfactory, a large fraction of the interval estimates would enclose the true value of θ. The fraction of such intervals enclosing θ is known as the _____ coefficient. This is not to be confused with the interval estimate, called the confidence _____.
The confidence coefficient actually gives the probability that a confidence interval will enclose θ.

interval estimates

confidence
interval

8.3 Point Estimation

In this section we will consider point estimation for the following parameters:
1. μ, the mean of a population of continuous measurements.
2. $(\mu_1 - \mu_2)$, the difference between the means for two populations of continuous measurements.
3. p, a binomial parameter.
4. $(p_1 - p_2)$, the difference in the parameters for two binomial populations.
We assume, in all four cases, that the samples are relatively large so that the estimators possess distributions in repeated sampling that are approximately normal due to the _____ _____ Theorem. The basic estimation problem is the same for all four cases and therefore we can discuss the problems in general by referring to the estimation of a parameter θ. Thus θ might be any one of the four parameters just mentioned.

Central Limit

To estimate the population parameter θ, a sample of size n $(y_1, y_2, \ldots, y_n)$ is randomly drawn from the population and an estimate of θ is calculated using $\hat{\theta}$. In repeated sampling, a distribution for $\hat{\theta}$ will be generated and will possess the following properties:
1. $E(\hat{\theta}) =$ _____.
2. $\hat{\theta}$ is approximately _____ distributed. Therefore, approximately 95% of the values of $\hat{\theta}$ will lie within two standard deviations of their mean, θ.

θ

normally

3. The symbol $\sigma_{\hat\theta}$ denotes the standard deviation of $\hat\theta$. Thus $\sigma_{\hat\theta}$ will be the standard deviation of the estimates generated by $\hat\theta$ in repeated sampling. The measure of goodness of a particular estimate is the distance that it lies from the target, θ. We call this distance _____ _____ _____ _____. Then when $\hat\theta$ possesses the properties stated above, the probability is approximately .95 that the error of estimation will be less than _____. We often refer to $2\sigma_{\hat\theta}$ as the *bound* on the error of estimation. By this we mean that the error will be less than $2\sigma_{\hat\theta}$ with high probability (say, near .95).

Complete the following table, filling in the estimator and its standard deviation where required:

Parameter	Estimator	Standard Deviation
μ	$\bar{y} = \sum_{i=1}^{n} y_i/n$	$\sigma/\sqrt{n}$
p	$\hat{p} = y/n$	_____
$\mu_1 - \mu_2$	_____	_____
$p_1 - p_2$	$\hat{p}_1 - \hat{p}_2$	$\sqrt{\dfrac{p_1 q_1}{n_1} + \dfrac{p_2 q_2}{n_2}}$

Notice that evaluation of the standard deviations given in the table may require values of parameters that are unknown. When the sample sizes are large, the sample estimates can be used to calculate an approximate standard deviation. As a rule of thumb, we will consider samples of size thirty or greater to be large samples.

Example 8.1
The mean length of stay for patients in a hospital must be known in order to estimate the number of beds required. The length of stay, recorded for a sample of 400 patients at a given hospital, gave a mean and standard deviation equal to 5.7 and 8.1 days, respectively. Give a point estimate for μ, the mean length of stay for patients entering the hospital, and place a bound of error on this estimate.

the error
of estimation

$2\sigma_{\hat\theta}$

$\sqrt{\dfrac{pq}{n}}$

$\bar{y}_1 - \bar{y}_2;\ \sqrt{\dfrac{\sigma_1^2}{n_1} + \dfrac{\sigma_2^2}{n_2}}$

Solution

1. The point estimate for μ is $\bar{y} =$ _____.

2. Since σ is unknown the *approximate* bound on error is

$$2\,\frac{s}{\sqrt{n}} = 2\left(\frac{8.1}{\sqrt{400}}\right) = \underline{\hspace{2cm}}.$$

5.7

.81

Self-Correcting Exercises 8A

1. In standardizing an examination, the average score on the exam must be known in order to differentiate among examinees taking the examination. The scores recorded for a sample of 93 examinees yielded a mean and standard deviation of 67.5 and 8.2, respectively. Estimate the true mean score for this examination and place bounds on the error of estimation.

2. Using the following data, give a point estimate with bounds on error for the difference in mortality rates in breast cancers where radical or simple mastectomy was used as a treatment:

	Radical	Simple
Number Died	31	41
Number Treated	204	191

3. A physician wishes to estimate the proportion of accidents on the California freeway system which result in fatal injuries to at least one person. He randomly checks the files on 50 automobile accidents and finds that 8 resulted in fatal injuries. Estimate the true proportion of fatal accidents, and place bounds on the error of estimation.

4. In measuring the tensile strength of two alloys, strips of the alloys were subjected to tensile stress and the force (measured in pounds) at which the strip broke recorded for each strip. The data is summarized below.

	Alloy 1	Alloy 2
$\bar{y}$	150.5	160.2
s^2	23.72	36.37
n	35	35

Use these data to estimate the true mean difference in tensile strength by finding a point estimate for $\mu_1 - \mu_2$ and placing a bound on the error of estimation.

8.4 Interval Estimation

An interval estimator is a rule that tells one how to calculate two points based on information contained in a sample. The objective is to form a narrow interval that will enclose the parameter. As in the case of point estimation,

confidence coefficient

θ

95

90

90

90

1.96

30

one can form many interval estimators (rules) for estimating the parameter of interest. Not all intervals generated by an interval estimator will actually enclose the parameter. The probability that an interval estimate will enclose the parameter is called the _____ _____.

Let $\hat{\theta}$ be an *unbiased* point estimator of θ and suppose that $\hat{\theta}$ generates a normal distribution of estimates in repeated sampling. The mean of this distribution of estimates is _____ and the standard deviation is $\sigma_{\hat{\theta}}$. Then _____% of the point estimates will lie within $1.96\sigma_{\hat{\theta}}$ of the parameter θ. Similarly, _____% will lie in the interval $\theta \pm 1.645\sigma_{\hat{\theta}}$.

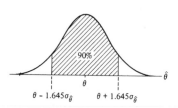

$$\theta - 1.645\sigma_{\hat{\theta}} \qquad \theta + 1.645\sigma_{\hat{\theta}}$$

Suppose that one were to construct an interval estimate by measuring the distance $1.645\sigma_{\hat{\theta}}$ on either side of $\hat{\theta}$. *Intervals constructed in this manner will enclose* θ _____% *of the time* (see below).

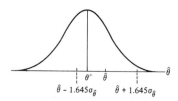

$$\hat{\theta} - 1.645\sigma_{\hat{\theta}} \qquad \hat{\theta} + 1.645\sigma_{\hat{\theta}}$$

Thus for a confidence interval with confidence coefficient $(1 - \alpha)$, we use

$$\left[\hat{\theta} \pm z_{\alpha/2}\sigma_{\hat{\theta}} \right]$$

to construct the interval estimate. The quantity $z_{\alpha/2}$ satisfies the relation $P[z > z_{\alpha/2}] = \alpha/2$ as indicated below:

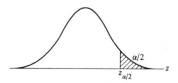

Not all good interval estimators are constructed by measuring $z_{\alpha/2}\sigma_{\hat{\theta}}$ on either side of the best point estimator, but this is true for the parameters μ, p, $(\mu_1 - \mu_2)$, and $(p_1 - p_2)$. These confidence intervals are good for samples that are large enough to achieve approximate normality for the distribution of $\hat{\theta}$ and good approximation for unknown parameters appearing in $\sigma_{\hat{\theta}}$.

A 95% confidence interval for μ is $\bar{y} \pm$ _____ $\sigma/\sqrt{n}$. As a rule of thumb, the sample size, n, must be greater than or equal to _____ in order that s be a good approximation to σ.

Give the z-values corresponding to the following confidence coefficients:

Confidence Coefficients	$z_{\alpha/2}$	
.95	1.96	1.645
.90	_____	2.58
.99	_____	

Example 8.2

To construct a 90% confidence interval for the mean length of hospital stay, μ, based on the sample of $n = 400$ patients ($\bar{y} = 5.7$ and $s = 8.1$) we calculate

$$\bar{y} \pm z_{\alpha/2}\sigma/\sqrt{n}.$$

Using $z_{.05} = 1.645$ and an estimate for σ given by $s = 8.1$, the interval estimate for the mean length of hospital stay is given as

$$5.7 \pm .67.$$

More properly, we estimate that _____ $< \mu <$ _____ with 90% confidence.   5.03; 6.37

The formula for a 95% confidence interval for a binomial parameter, p, is

 _____ .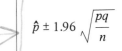

$$\hat{p} \pm 1.96\sqrt{\frac{pq}{n}}$$

Note that $\hat{p}$ is used to approximate p in the formula for $\sigma_{\hat{p}}$ since its value is unknown.

Example 8.3

An experimental rehabilitation technique employed on released convicts showed that 79 of a total of 121 men subjected to the technique pursued useful and crime-free lives for a three-year period following prison release. Find a 95% confidence interval for p, the probability that a convict subjected to the rehabilitation technique will follow a crime-free existence for at least three years after prison release.

Solution

The sampling described above satisfies the requirements of a binomial experiment consisting of $n = 121$ trials. In estimating the parameter p with a 95% confidence interval we use the estimator

$$\hat{p} \pm \underline{\hspace{1.5cm}}\sqrt{\frac{pq}{n}}.$$ 1.96

Since p is unknown, the sample value, $\hat{p}$, will be used in the approximation of $\sqrt{pq/n}$. Collecting pertinent information, we have

1. $$\hat{p} = \frac{y}{n} = \frac{79}{121} = \underline{\hspace{1.5cm}}.$$ .65

2. $\sqrt{\dfrac{\hat{p}\hat{q}}{n}} = \sqrt{\dfrac{(.65)(.35)}{121}} = .04.$

3. The interval estimate is given as

$$.65 \pm 1.96(.04)$$

.08

or $.65 \pm$ _____ .

.57; .73

4. We estimate that _____ $< p <$ _____ with 95% confidence.

The exact values for standard deviations of estimators cannot usually be found because they are functions of unknown population parameters. For the following estimators, give the standard deviation and the best approximation of the standard deviation for use in confidence intervals.

Estimator ($\hat{\theta}$)	Standard Deviation ($\sigma_{\hat{\theta}}$)	Best Approximation of Standard Deviation ($\hat{\sigma}_{\hat{\theta}}$)
$\bar{y}$	$\sigma/\sqrt{n}$	_____
$\hat{p}$	$\sqrt{\dfrac{pq}{n}}$	_____
$\bar{y}_1 - \bar{y}_2$	$\sqrt{\dfrac{\sigma_1^2}{n_1} + \dfrac{\sigma_2^2}{n_2}}$	$\sqrt{\dfrac{s_1^2}{n_1} + \dfrac{s_2^2}{n_2}}$
$\hat{p}_1 - \hat{p}_2$	$\sqrt{\dfrac{p_1 q_1}{n_1} + \dfrac{p_2 q_2}{n_2}}$	$\sqrt{\dfrac{\hat{p}_1 \hat{q}_1}{n_1} + \dfrac{\hat{p}_2 \hat{q}_2}{n_2}}$

(left margin answers: $s/\sqrt{n}$; $\sqrt{\dfrac{\hat{p}\hat{q}}{n}}$)

The large sample confidence intervals for μ, p, $\mu_1 - \mu_2$, and $p_1 - p_2$ will be

$$\hat{\theta} \pm z_{\alpha/2}\, \hat{\sigma}_{\hat{\theta}}$$

where $\hat{\theta}$ is given by $\bar{y}$, $\hat{p}$, $\bar{y}_1 - \bar{y}_2$, and $\hat{p}_1 - \hat{p}_2$, respectively. The above table will determine the appropriate formula for $\sigma_{\hat{\theta}}$ or $\hat{\sigma}_{\hat{\theta}}$.

Example 8.4

An experiment was conducted to compare the mean absorptions of drug in specimens of muscle tissue. Seventy-two tissue specimens were randomly divided between two drugs A and B with 36 assigned to each drug, and the drug absorption was measured for the 72 specimens. The means and variances for the two samples were, $\bar{y}_1 = 7.8$, $s_1^2 = .10$ and $\bar{y}_2 = 8.4$, $s_2^2 = .06$, respectively. Find a 95% confidence interval for the difference in mean absorption rates.

Solution

We are interested in placing a confidence interval about the parameter, _____ . The confidence interval is

(left margin: $\mu_1 - \mu_2$)

$$(\bar{y}_1 - \bar{y}_2) \pm z_{\alpha/2} \sqrt{\frac{s_1^2}{n_1} + \frac{s_2^2}{n_2}}$$

$$(7.8 - 8.4) \pm \underline{\hspace{1cm}} \sqrt{\frac{.10}{36} + \frac{.06}{36}}$$ 1.96

$$\underline{\hspace{1cm}} \pm .131$$ −.6

or $\underline{\hspace{1cm}} < \mu_1 - \mu_2 < \underline{\hspace{1cm}}.$ −.731; −.469

Example 8.5

The voting records at two precincts were compared based on samples of 400 voters each. Those voting Democratic numbered 209 and 263, respectively. Estimate the difference in the fraction voting Democratic for the two precincts using a 90% confidence interval.

Solution

The confidence interval is

$$(\hat{p}_1 - \hat{p}_2) \pm z_{\alpha/2} \sqrt{\frac{\hat{p}_1 \hat{q}_1}{n_1} + \frac{\hat{p}_2 \hat{q}_2}{n_2}}$$

$$(.5225 - .6575) \pm \underline{\hspace{1cm}} \sqrt{\frac{(.5225)(.4775)}{400} + \frac{(.6575)(.3425)}{400}}$$ 1.645

$$\underline{\hspace{1cm}} \pm .057$$ −.135

or $\underline{\hspace{1cm}} < p_1 - p_2 < \underline{\hspace{1cm}}.$ −.192; −.078

Self-Correcting Exercises 8B

1. Suppose it is necessary to estimate the percentage of students on the University of Florida campus who favor constitutional revision of the State Constitution. In a random sample of 65 students, 30 stated that they were in favor of revision. Estimate the percentage of students favoring revision with a 98% confidence interval.
2. It is wished to estimate the difference in accident rates between youth and adult drivers. The following data were collected from two random samples, where y is the number of drivers that were involved in one or more accidents:

Youths	Adults
$n_1 = 100$	$n_2 = 200$
$y_1 = 50$	$y_2 = 60$

Estimate the difference, $(p_1 - p_2)$, with a 95% confidence interval.

3. The yearly incomes of high school teachers in two cities yielded the following tabulation:

	City 1	City 2
Number of Teachers	90	60
Average Income	12,520	11,210
Standard Deviation	1,510	950

a. If the teachers from each city are thought of as samples from two populations of high school teachers, use the data to construct a 99% confidence interval for the difference in mean annual incomes.

b. Using the results of a. would you be willing to conclude that these two city schools belong to populations having the same mean annual income?

4. A sample of 39 cigarettes of a certain brand, tested for nicotine content, gave a mean of 22 and a standard deviation of 4 milligrams. Find a 90% confidence interval for the average nicotine content of this brand of cigarette.

8.5 Choosing the Sample Size

One of the first steps in planning an experiment is deciding on the quantity of information that we wish to buy. At first glance it would seem difficult to specify a measure of the quantity of information in a sample relevant to a parameter of interest. However, such a practical measure is available in the bound on the error of estimation; or, alternatively, we could use the half-width of the confidence interval for the parameter.

The larger the sample size, the greater will be the amount of information contained in the sample. This intuitively appealing fact is evident upon examination of the large sample confidence intervals. The width of each of the four confidence intervals described in the preceding section is inversely proportional to the square root of the _____ _____.

Suppose that $\hat{\theta}$ is an estimator of θ and satisfies the conditions for the large sample estimators previously discussed. Then the bound, B, on the error of estimation will be $2\sigma_{\hat{\theta}}$. This means that the error (in repeated sampling) will be less than $2\sigma_{\hat{\theta}}$ with probability _____. If B represents the desired bound on the error, then for a _point_ estimator $\hat{\theta}$ the restriction is:

a. $2\sigma_{\hat{\theta}} = B$.

In an interval estimation problem with $(1 - \alpha)$ confidence coefficient, the restriction is:

b. $z_{\alpha/2}\sigma_{\hat{\theta}} = B$.

For all practical purposes, a. and b. are equivalent for a confidence coefficient equal to .95.

Example 8.6

Suppose it is known that $\sigma = 2.25$ and it is desired to estimate μ with a bound on the error of estimation less than or equal to 0.5 units with probability .95. How large a sample should be taken?

sample size

.95

Solution
The estimator for μ is $\bar{y}$ with standard deviation $\sigma/\sqrt{n}$; $1 - \alpha = .95$, $\alpha/2 = .025$, $z_{.025} = 1.96$, or approximately 2. Hence we solve

$$2(\sigma/\sqrt{n}) = B$$

or $$2(2.25/\sqrt{n}) = 0.5$$

$$2(2.25/0.5) = \sqrt{n}$$

$$\underline{\hspace{2cm}} = \sqrt{n} \qquad\qquad 9$$

$$\underline{\hspace{2cm}} = n. \qquad\qquad 81$$

The solution is to take a sample of size _____ or greater to insure that   81
the bound is less than or equal to 0.5 units. Had we wished to have the same
bound with probability .99, the value $z_{.005} = $ _____ would have been   2.58
used, resulting in the following solution:

$$\frac{2.25}{\sqrt{n}} = 0.5 \qquad\qquad 2.58$$

$$\sqrt{n} = \frac{2.58\,(2.25)}{0.5}$$

$$n = \underline{\hspace{2cm}}. \qquad\qquad 134.79$$

Hence, a sample of size _____ or greater would be taken to insure   135
estimation with $B = 0.5$ units.

Example 8.7
If an experimenter wished to estimate the fraction of university students
that daily read the college newspaper correct to within .02 with probability
.90, how large a sample of students should he take?

Solution

To estimate binomial p with a 90% confidence interval we would use

$$\hat{p} \pm \underline{\hspace{2cm}} \sqrt{\frac{pq}{n}}. \qquad\qquad 1.645$$

We wish to find a sample size n so that

$$1.645\sqrt{\frac{pq}{n}} = .02.$$

Since neither p nor $\hat{p}$ is known, we can solve for n by assuming the worst
possible variation, which occurs when $p = q = $ _____. Hence we solve   0.5

$$1.645 \sqrt{\frac{(.5)(.5)}{n}} = .02$$

$$\frac{1.645(.5)}{.02} = \sqrt{n}$$

1691.27

or $\underline{\hspace{2cm}} = n.$

1692

Therefore, we should take a sample of size $\underline{\hspace{2cm}}$ or greater to achieve the required bounds even if faced with the maximum variation possible.

Example 8.8
An experiment is to be conducted to compare two different sales techniques at a number of sales centers. Suppose that the range of sales for the sales centers is expected to be $4000. How many centers should be included for each of the sales techniques in order to estimate the difference in mean sales correct to within $500?

Solution
We will assume that the two sample sizes are equal, that is, $n_1 = n_2 = n$, and that the desired confidence coefficient is .95. Then

$2\sqrt{\dfrac{\sigma_1^2}{n} + \dfrac{\sigma_2^2}{n}}$

$$\underline{\hspace{2cm}} = B.$$

1000

The quantities σ_1^2 and σ_2^2 are unknown but we know that the range is expected to be $4000. Then we would take $\sigma_1 = \sigma_2 = \underline{\hspace{2cm}}$ as the best available approximation. Then, substituting into the above equation,

$\dfrac{2(1000)^2}{n}$

$$2\sqrt{\underline{\hspace{1.5cm}}} = 500$$

32

or $\qquad n = \underline{\hspace{2cm}}.$

32

Thus $n = \underline{\hspace{2cm}}$ sales centers would be required for each of the two sales techniques.

Self-Correcting Exercises 8C

1. A device is known to produce measurements whose errors in measurement are normally distributed with a standard deviation $\sigma = 8$ mm. If the average measurement is to be reported, how many repeated measurements should be used so that the error in measurement is no larger than 3 mm. with probability .95?

2. An experiment is to be conducted to compare the taste threshold levels for each of two food additives as measured by their concentrations in parts per million. How many subjects should be included in each experimental group in order to estimate the mean difference in threshold levels to within 10 units if the range of the measurements is expected to be approximately 80 ppm. for both groups?

3. How many individuals from each of two politically oriented groups should be included in a poll designed to estimate the true difference in proportions favoring a tuition increase at the state university correct to within .01 with probability .95? (In the absence of any prior information regarding the values of p_1 and p_2, solve the problem assuming maximum variation.)

4. If a mental health agency would like to estimate the percentage of local clinic patients that are referred to their counseling center to within 5 percentage points with 90% accuracy, how many patient records should be sampled?

8.6 A Statistical Test of an Hypothesis

We now leave estimation and turn our attention to a decision-making form of inference, hypothesis-testing. In hypothesis-testing, we formulate an hypothesis about a population in terms of its _____ and then after observing a _____ drawn from this population, decide whether our sample value could have come from the hypothesized population. We then accept or reject the hypothesized value.

    parameters

    sample

    A statistical test of an hypothesis consists of four parts:

1. _____ _____: (H_0) This is the hypothesis to be tested and gives hypothesized values for one or more population parameters.

    Null hypothesis

2. _____ _____: (H_a) This is the hypothesis against which H_0 is tested. We look for evidence in the sample that will cause us to reject H_0 in favor of H_a.

    Alternative hypothesis

3. Test statistic: This function of the sample values extracts the information about the parameter contained in the sample. The observed value of the test statistic leads us to reject one hypothesis and accept the other.

4. Rejection region: Once the test statistic to be used is selected, the entire set of values that the statistic may assume is divided into two regions. The acceptance region consists of those values most likely to have arisen if H_0 were true. The rejection region consists of those values most likely to have arisen if H_a were true. If the observed value of the test statistic falls in the rejection region, H_0 is rejected; if it falls in the acceptance region, H_0 is accepted.

The statistical test described above may result in one of two types of error:

1. Type I error: Rejecting H_0 when H_0 is true.
2. Type II error: Accepting H_0 when H_a is true.

We define

1. $\alpha = P[\text{Type I error}]$.

2. $\beta = P[\text{Type II error}]$.

For a fixed sample size n,

1. if the rejection region is enlarged, α will (increase, decrease) while β will (increase, decrease);

    increase

    decrease

decrease
increase
decrease

2. if the rejection region is made smaller, α will (increase, decrease) while β will (increase, decrease).

When the sample size n is increased, both α and β will (increase, decrease) because of the added information contained in the sample.

Notice that β is a function of H_a, since by definition,

$$\beta = P[\text{accepting } H_0 | H_a \text{ true}].$$

By saying H_a is true, we mean that the true value of the population parameter is that given by H_a, and β is computed using that value.

8.7 A Large Sample Test of an Hypothesis

As in previous sections, the parameter of interest $(\mu, \mu_1 - \mu_2, p, \text{ or } p_1 - p_2)$ will be referred to as θ. If an *unbiased* point estimator, $\hat{\theta}$, exists for θ and if $\hat{\theta}$ is normally distributed, we can employ $\hat{\theta}$ as a test statistic to test the hypothesis, $H_0: \theta = \theta_0$.

larger

If H_a states that $\theta > \theta_0$, that is, the value of the parameter is greater than that given by H_0, then the sample value for $\hat{\theta}$ should reflect this fact and be (larger, smaller) than a value of $\hat{\theta}$ when sampling from a population whose mean is θ_0. Hence we would reject H_0 for large values of $\hat{\theta}$. Large can be interpreted as too many standard deviations to the right of the mean, θ_0. The value of $\hat{\theta}$ selected to separate the acceptance and rejection regions is called

critical value

the _____ _____ of the test statistic.

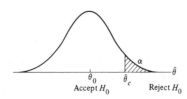

θ_0
Accept H_0 $\hat{\theta}_c$ Reject H_0

α

$\hat{\theta}_c$ in the diagram represents the critical value of $\hat{\theta}$ and the shaded area to the right of $\hat{\theta}_c$ is equal to _____. This is a one-tailed statistical test.

A similar picture could have been used with the critical value of $\hat{\theta}$ to the

$<$
one-tailed

left of the mean for testing $H_0: \theta = \theta_0$ against $H_a: \theta$ _____ θ_0. Then we would reject for values of $\hat{\theta}$ lying too many standard deviations to the left of θ_0 (resulting in a _____ - _____ test in the left tail) and would reject H_0 for small values of $\hat{\theta}$.

two

A third type of alternative hypothesis would be $H_a: \theta \neq \theta_0$ where we seek departures either greater or less than θ_0. This results in a _____ - tailed statistical test.

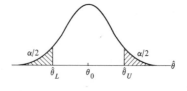

$\alpha/2$ $\alpha/2$
$\hat{\theta}_L$ θ_0 $\hat{\theta}_U$

In order that the probability of a Type I error be equal to α, two critical values of $\hat{\theta}$ must be found, one having area $\alpha/2$ to its right ($\hat{\theta}_U$) and one having area $\alpha/2$ to its left ($\hat{\theta}_L$). H_0 will be rejected if $\hat{\theta} \geqslant \hat{\theta}_U$ or $\hat{\theta} \leqslant \hat{\theta}_L$.

Since the estimator $\hat{\theta}$ is normally distributed, we can standardize the normal variable, $\hat{\theta}$, by converting the distance that $\hat{\theta}$ departs from θ_0 to z (the number of standard deviations to the left or right of the mean). Thus, we will use z as the test statistic. The four elements of the test are:

1. $H_0: \theta = \theta_0$.

2. One of the three alternatives:

a. $H_a: \theta > \theta_0$ (right-tailed).

b. $H_a: \theta < \theta_0$ (left-tailed).

c. $H_a: \theta \neq \theta_0$ (two-tailed).

3. Test statistic: z, where

$$z = \frac{\hat{\theta} - \theta_0}{\sigma_{\hat{\theta}}}.$$

4. Rejection region:
   a. For $H_a: \theta > \theta_0$

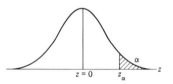

Reject H_0 if $z > z_\alpha$

b. For $H_a: \theta < \theta_0$

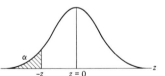

Reject H_0 if $z < -z_\alpha$

c. For $H_a: \theta \neq \theta_0$

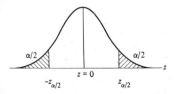

Reject H_0 if $z > z_{\alpha/2}$
or $z < -z_{\alpha/2}$
($|z| > z_{\alpha/2}$)

We now apply this test of an hypothesis.

Example 8.9
Test of a population mean μ. Test the hypothesis at the $\alpha = .05$ level that a population mean $\mu = 10$ against the hypothesis that $\mu > 10$ if for a sample of 81 observations, $\bar{y} = 12$ and $s = 3.2$. Note that

a. $\theta = \mu$.   d. $\sigma_{\hat{\theta}} = \sigma/\sqrt{n}$.

b. $\theta_0 = \mu_0 = 10$.   e. $\alpha = .05$.

c. $\hat{\theta} = \bar{y}$.

Solution
Since σ is unknown, use s, the sample standard deviation, as its approximation. Then, the elements of the test are:

1. $H_0 : \mu = 10$.

2. $H_a : \mu > 10$.

3. Test statistic

$$z = \frac{\bar{y} - \mu_0}{\sigma/\sqrt{n}} . \qquad \text{(using s if σ is unknown.)}$$

4. Rejection region:

1.645

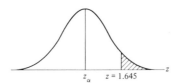

Reject H_0 if $z > $ _____

Having defined the test, calculate z.

12
3.2

$$z = \frac{\bar{y} - \mu_0}{s/\sqrt{n}} = \frac{\underline{\hspace{1cm}} - 10}{\underline{\hspace{1cm}}/\sqrt{81}}$$

$z = 2/.356 = 5.62$.

reject

Since $z = 5.62 > 1.645$, the decision is (reject, do not reject) H_0 with $\alpha = .05$.

Example 8.10
Test of a binomial p. In assessing the effect of the color of food on taste preferences, 65 subjects were each asked to taste two samples of mashed potatoes, one of which was colored pink, the other its natural color. Although both samples were identical except for the pink color produced by the addition of a drop of tasteless food coloring, 53 of the subjects preferred the taste of the sample possessing its natural color. Does this indicate that subjects tend to be adversely affected by the pink color? Test at the $\alpha = .01$ level of significance.

Solution

In this problem, with $p = P[choose\ natural\ color]$.

a. $\theta = p$.

d. $\sigma_{\hat\theta} = \sqrt{\dfrac{p_0 q_0}{n}}$.

b. $\theta_0 = p_0 = \underline{\quad\quad}$.

½

e. $\alpha = \underline{\quad\quad}$.

.01

c. $\hat\theta = \hat p = y/n$

Note that p_0 and q_0 are used in $\sigma_{\hat p}$ since we are testing $H_0: p = p_0$. The elements of the test are

1. $H_0: p = \frac{1}{2}$.

2. $H_a: p > \frac{1}{2}$.

3. Test statistic:

$$z = \frac{\hat p - p_0}{\sqrt{\dfrac{p_0 q_0}{n}}}.$$

4. Rejection region:

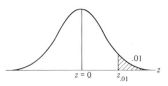

Reject H_0 if $z > \underline{\quad\quad}$.

2.33

$z = 0$ $z_{.01}$

To calculate z, we need $\hat p = y/n = 53/65 = .815$.

$$z = \frac{.815 - \underline{\quad\quad}}{\sqrt{\dfrac{.5(.5)}{65}}} = \frac{.315}{.062} = 5.08.$$

.500

Since $z = 5.08 > 2.33$ we (reject, do not reject) H_0 and conclude that color (does, does not) affect the choice of potato.

reject
does

Example 8.11

Test of an hypothesis concerning $(\mu_1 - \mu_2)$. Suppose an educator is interested in testing whether a new teaching technique is superior to an old teaching technique. The criterion will be a test given at the end of a six-week period and the technique resulting in a significantly higher score will be judged superior. Test the hypothesis that the test means for the techniques are the same against the alternative hypothesis that the new technique is superior at the $\alpha = .05$ level, based on the following sample data:

	New Technique	Old Technique
	$\bar{y}_1 = 69.2$	$\bar{y}_2 = 67.5$
	$s_1^2 = 49.3$	$s_2^2 = 64.5$
	$n_1 = 50$	$n_2 = 80$

Solution

0

1. $H_0: \mu_1 - \mu_2 = \underline{\hspace{2cm}}.$

2. $H_a: \mu_1 - \mu_2 > 0.$

3. Test statistic:

$$z = \frac{(\bar{y}_1 - \bar{y}_2) - 0}{\sqrt{\dfrac{s_1^2}{n_1} + \dfrac{s_2^2}{n_2}}}$$

(since σ_1^2 and σ_2^2 are unknown).

4. Rejection region:

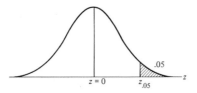

Reject H_0 if $z > \underline{\hspace{2cm}}$

1.645

Computing the value of the test statistic,

69.2; 67.5

$$z = \frac{(\underline{\hspace{1.5cm}} - \underline{\hspace{1.5cm}}) - 0}{\sqrt{\dfrac{49.3}{50} + \dfrac{64.5}{80}}}$$

1.7; 1.27

$$= \frac{1.7}{\sqrt{\dfrac{716.9}{400}}} = \frac{1.7}{1.34} = \underline{\hspace{2cm}}.$$

does not; will not

Since the value of z (does, does not) fall in the rejection region, we (will, will not) reject H_0. Before deciding to accept H_0 as true, we may wish to evaluate the probability of a Type II error for meaningful values of $\mu_1 - \mu_2$ described by H_a. Until this is done, we shall state our decision as "Do not reject H_0."

Example 8.12
To investigate possible differences in attitude about a current state political problem, 100 randomly selected voters between the ages of 18 and 25 were polled, and 100 randomly selected voters over age 25 were polled. Each was

asked if he agreed with the government's position on the problem. Forty-five of the first group agreed, while 63% of the second group agreed. Do these data represent a significant difference in attitude for these two groups?

Solution

This problem involves a test of the difference between two binomial proportions, $p_1 - p_2$. The relevant data are:

	Group 1	Group 2
n	100	100
$\hat{p}$	.45	.63

1. $H_0: p_1 - p_2$ _____. $= 0$

2. $H_a: p_1 - p_2$ _____. $\neq 0$

3. For testing the hypothesis of *no difference* between proportions, the test statistic is

$$z = \frac{(\hat{p}_1 - \hat{p}_2) - 0}{\sqrt{\hat{p}\hat{q}\left(\dfrac{1}{n_1} + \dfrac{1}{n_2}\right)}}$$

with

$$\hat{p} = \frac{y_1 + y_2}{n_1 + n_2}.$$

4. Rejection region:

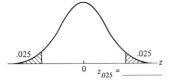

$z_{.025} =$ _____   1.96

For a two-tailed test with $\alpha = .05$, we shall reject H_0 if $|z| >$ _____.   1.96
To calculate the test statistic, we need

$$\hat{p} = \frac{y_1 + y_2}{n_1 + n_2} = \frac{45 + 63}{200} = \text{_____}.$$ .54

Then

$$z = \frac{(.45 - .63) - 0}{\sqrt{(.54)(.46)(2/100)}}$$

$$= \frac{\text{_____}}{.0705}$$ $-.18$

$$= \text{_____}.$$ -2.55

reject
is

Since $|-2.55| = 2.55 > 1.96$, we (reject, do not reject) H_0 and conclude that there (is, is not) a significant difference in opinion between these two age groups with respect to this issue.

Self-Correcting Exercises 8D

1. To investigate a possible "built-in" sex-bias in a graduate school entrance examination, 50 male and 50 female graduate students who were rated as above-average graduate students by their professors were selected to participate in the study by actually taking this test. Their test results on this examination are summarized below.

	Males	Females
$\bar{y}$	720	693
s^2	104	85
n	50	50

Do these data indicate that males will on the average score higher than females of the same ability on this exam? Use $\alpha = .05$.

2. A machine shop is interested in determining a measure of the current year's sales revenue in order to compare it with known results from last year. From the 9,682 sales invoices for the current year to date, the management randomly selected $n = 400$ invoices and from each recorded y, the sales revenue per invoice. Using the following data summary, test the hypothesis that the mean revenue per invoice is $3.35, the same as last year, versus the alternative hypothesis that the mean revenue per invoice is different from $3.35 with $\alpha = .05$.

Data Summary

$n = 400$

$$\sum_{i=1}^{400} y_i = \$1264.40$$

$$\sum_{i=1}^{400} y_i^2 = 4970.3282.$$

3. A physician found 480 men and 420 women among 900 patients admitted to a hospital with a certain disease. Is this consistent with the hypothesis that in the population of patients hospitalized with this disease, half the cases are male? Use $\alpha = .10$.

4. Random samples of 100 shoes manufactured by machine A and 50 shoes manufactured by machine B showed 16 and 6 defective shoes, respectively. Do these data present sufficient evidence to suggest a difference in the performances of the machines? Use $\alpha = .05$.

EXERCISES

1. List the two essential elements of any inference-making procedure.
2. What are two desirable properties of a point estimator, $\hat{\theta}$?
3. A bank was interested in estimating the average size of its savings accounts for a particular class of customer. If a random sample of 400 such accounts showed an average amount of $61.23 and a standard deviation of $18.20, place 90% confidence limits on the actual average account size.
4. If 36 measurements of the specific gravity of aluminum had a mean of 2.705 and a standard deviation of 0.028, construct a 98% confidence interval for the actual specific gravity of aluminum.
5. An appliance dealer sells toasters of two different brands, brand A and brand B. Let p_1 denote the fraction of brand A toasters which are returned to him by customers as defective, and let p_2 represent the fraction of brand B toasters which are rejected by customers as defective. Suppose that of 200 brand A toasters sold, 14 were returned as defective, while of 450 brand B toasters sold, 18 were returned as defective. Provide a 90% confidence interval for $p_1 - p_2$.
6. What are the four essential elements of a statistical test of an hypothesis?
7. Assume that a certain set of "early returns" in an election is actually a random sample of size 400 from the voters in that election. If 225 of the voters in the sample voted for candidate A, could we assert with $\alpha = .01$ that candidate A has won?
8. In a sample of 400 seeds, 240 germinated. Estimate the true germination percentage with a 95% confidence interval.
9. A doctor wishes to estimate the average nicotine content in a certain brand of cigarettes correct to within 0.5 milligram. From previous experiments it is known that σ is in the neighborhood of 4 milligrams. How large a sample should the doctor take to be 95% confident of his estimate?
10. A manufacturer of dresses believes that approximately 20% of his product contains flaws. If he wishes to estimate the true percentage to within 8%, how large a sample should he take?
11. It is desired to estimate $\mu_1 - \mu_2$ from information contained in independent random samples from populations with variances $\sigma_1^2 = 9$ and $\sigma_2^2 = 16$. If the two sample sizes are to be equal $(n_1 = n_2 = n)$, how large should n be in order to estimate $\mu_1 - \mu_2$ with an error less than 1.0 (with probability equal to .95)?
12. To compare the effect of stress in the form of noise upon the ability to perform a simple task, 70 subjects were divided into two groups, the first group of 30 subjects were to act as a control while the second group of 40 were to be the experimental group. Although each subject performed the task in the same control room, each of the experimental group subjects had to perform the task while loud rock music was being played in the room. The time to finish the task was recorded for each subject and the following summary was obtained:

	Control	Experimental
n	30	40
$\bar{y}$	15 minutes	23 minutes
s	4 minutes	10 minutes

Find a 99% confidence interval for the difference in mean completion times for these two groups.

13. A grocery store operator claims that the average waiting time at a checkout counter is 3.75 minutes. To test this claim, a random sample of 30 observations was taken. Test the operator's claim at the 5% level of significance using the sample data shown below:

Waiting Time in Minutes				
3	4	3	4	1
1	0	5	3	2
4	3	1	2	0
3	2	0	3	4
1	3	2	1	3
2	4	2	5	2

14. In order to test the effectiveness of a vaccine, 150 experimental animals were given the vaccine; 150 were not. All 300 were then infected with the disease. Among those vaccinated, ten died as a result of the disease. Among the control group (i.e., those not vaccinated), there were 30 deaths. Can we conclude that the vaccine is effective in reducing the mortality rate? Use a significance level of .025.

15. Two diets were to be compared. Seventy-five individuals were selected at random from a population of overweight people. Forty of this group were assigned diet A and the other thirty-five were placed on diet B. The weight losses in pounds over a period of one week were found and the following quantities recorded:

	Sample Size	Sample Mean (lbs)	Sample Variance
Diet A	40	10.3	7
Diet B	35	7.3	3.25

a. Do these data allow the conclusion that the expected weight loss under diet A (μ_A) is greater than the expected weight loss under diet B (μ_B)? Test at the .01 level. Draw the appropriate conclusion.

b. Construct a 90% confidence interval for $\mu_A - \mu_B$.

Chapter   9

INFERENCE FROM SMALL SAMPLES

9.1   Introduction

Large sample methods for making inferences about a population were considered in the preceding chapter. When the sample size was large, the _____ _____ Theorem assured the approximate normality of the distribution of the estimators $\bar{y}$ or $\hat{p}$. However, time, cost, or other limitations may prevent an investigator from collecting enough data to feel confident in using large sample techniques. When the sample size is small, $n < 30$, the Central Limit Theorem may no longer apply. This difficulty can be overcome if the investigator is reasonably sure that his measurements constitute a sample from a _____ population.

Central Limit

normal

   The results presented in this chapter are based upon the assumption that the observations being analyzed have been _____ drawn from a normal population. This assumption is not as restrictive as it sounds, since the normal distribution can be used as a model in cases where the underlying distribution is mound-shaped and fairly symmetrical.

randomly

9.2   Student's *t* Distribution

When the sample size is large, the statistic

$$\frac{\bar{y} - \mu}{\sigma/\sqrt{n}}$$

is approximately distributed as the standard normal random variable z. What can be said about this statistic when n, the sample size, is small and the sample variance s^2 is used to estimate σ^2?

   If the parent population is not normal (nor approximately normal) the behavior of the statistic given above is not known in general when n is small. Its distribution could be empirically generated by repeated sampling from the population of interest. If the parent population *is* normal we can rely upon the results of W. S. Gosset, who published under the pen name Student. He drew repeated samples from a normal population and tabulated the distribution of the statistic that he called t, where

$$t = \frac{\bar{y} - \mu}{s/\sqrt{n}} \; .$$

The resulting distribution for t had the following properties:

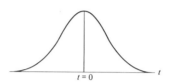

mound	1. The distribution is _____-shaped.
symmetrical	2. The distribution is _____ about the value $t = 0$.
more	3. The distribution has more flaring tails than z; hence t is (more, less) variable than the z statistic.
n	4. The shape of the distribution changes as the value of _____, the sample size, changes.
	5. As the sample size, n, becomes large, the t distribution becomes identical
standard normal	to the _____ _____ distribution.

These results are based on the following two assumptions:

normal	1. The parent population has a _____ distribution. The t-statistic
mound	is, however, relatively stable for non-normal _____-shaped distributions.
random	2. The sample is a _____ sample. When the population is normal, this assures us that $\bar{y}$ and s^2 are independent.

For a fixed sample size, n, the statistic

$$z = \frac{\bar{y} - \mu}{\sigma/\sqrt{n}}$$

| one; $\bar{y}$ | contains exactly _____ random quantity, the sample mean, _____. |

However, the statistic

$$t = \frac{\bar{y} - \mu}{s/\sqrt{n}}$$

two; $\bar{y}$; s	contains _____ random quantities, _____ and _____.
more	This accounts for the fact that t is (more, less) variable than z. In fact $\bar{y}$ may be large while s is small or $\bar{y}$ may be small while s is large. Hence it is said
independent	that $\bar{y}$ and s are _____, which means that the value assumed by $\bar{y}$ in no way determines the value of s.

As the sample size changes, the corresponding t distribution changes so that each value of n determines a different probability distribution. This is due to the variability of s^2, which appears in the denominator of t. Large sample

| more | sizes produce (more, less) stable estimates of σ^2 than do small sample sizes. These different probability curves are identified by the degrees of freedom associated with the estimator of σ^2. |

The term degrees of freedom can be explained in the following way. The sample estimate, s^2, uses the sum of squared deviations in its calculation.

Recall that $\sum\limits_{i=1}^{n} (y_i - \bar{y}) =$ _____ . This means that if we know the | 0

values of $n - 1$ deviations, we can determine the last value uniquely since their sum must be zero. Therefore, the sum of squared deviations,

$\sum\limits_{i=1}^{n} (y_i - \bar{y})^2$ contains only _____ independent deviations, and not n | $n - 1$

independent deviations as one might expect. Degrees of freedom refer to the number of independent deviations that are available for estimating σ^2. When n observations are drawn from one population, we use

$$\hat{\sigma}^2 = s^2 = \frac{\sum\limits_{i=1}^{n} (y_i - \bar{y})^2}{n - 1} .$$

In this case, the degrees of freedom for estimating σ^2 are _____ and the | $n - 1$
resulting t distribution is indexed as having _____ degrees of freedom. | $n - 1$

The Use of Tables for the t Distribution

We define t_α as that value of t having an area equal to α to its _____ | right
and $-t_\alpha$ is that value of t having an area equal to α to its _____ . See | left
the following diagram:

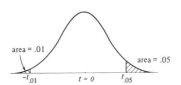

area = .01

area = .05

$-t_{.01}$ $t = 0$ $t_{.05}$

The distribution of t is _____ about the value $t = 0$; hence, only the | symmetrical
positive values of t need be tabulated. Problems involving left-tailed values of
t can be solved in terms of right-tailed values, as was done with the z-statistic.
A negative value of t simply indicates that you are working in the (left, right) | left
tail of the distribution.

   Table 4 of your text tabulates *commonly used* critical values, t_α, based on
$1, 2, \ldots, 29, \infty$ degrees of freedom for $\alpha = .100, .050, .025, .010, .005$. Along
the top margin of the table you will find columns labeled t_α for the various
values of α, while along the right margin you will find a column marked
degrees of freedom, *d.f.* By cross-indexing you can find the value t having an
area equal to α to its right and having the proper degrees of freedom.

Example 9.1

To find the critical value of t for $\alpha = .05$ with five degrees of freedom, find
five in the right margin. Now by reading across, you will find $t = 2.015$ in the
$t_{.05}$ column. In the same manner, we find that for 12 degrees of freedom,

2.179

$t_{.025} =$ _____ . In using Table 4, a student should think of his problem in terms of α, the area to the right of the value of t, and the degrees of freedom used to estimate σ^2. The reader is asked to compare the different values of t based on an infinite number of degrees of freedom with those for a corresponding z. One can perhaps see the reason for choosing a sample size

30

σ

greater than _____ as the dividing point for using the z distribution when the standard deviation s is used as an estimate for _____.

Example 9.2

Find the critical values for t when t_α is that value of t with an area of α to its right, based on the following degrees of freedom:

		α	d.f.	t
2.920	a.	.05	2	_____
3.169	b.	.005	10	_____
1.313	c.	.10	28	_____
2.583	d.	.01	16	_____
2.086	e.	.025	20	_____

Students taking their first course in statistics usually ask the following questions at this point: "How will I know whether I should use z or t? Is sample size the only criterion I should apply?" No, sample size is not the only criterion to be used. When the sample size is *large*, both

$$T_1 = \frac{\bar{y} - \mu}{\sigma/\sqrt{n}} \quad \text{and} \quad T_2 = \frac{\bar{y} - \mu}{s/\sqrt{n}}$$

behave as a standard normal random variable z, regardless of the distribution of the parent population. When the sample size is *small* and the sampled population is *not normal*, then in general, neither T_1 nor T_2 behaves as z or t. In the special case when the parent population is *normal*, then T_1 behaves as z and T_2 behaves as t.

Use this information to complete the following table when the sample is drawn from a *normal* distribution:

	Sample Size	
Statistic	$n < 30$	$n \geqslant 30$
$\dfrac{\bar{y} - \mu}{s/\sqrt{n}}$	_____	t or app. z
$\dfrac{\bar{y} - \mu}{\sigma/\sqrt{n}}$	_____	_____

t

z; z

9.3 Small Sample Inferences Concerning a Polulation Mean

Small Sample Test Concerning a Population Mean, μ

A test of an hypothesis concerning the mean, μ, of a *normal* population when $n < 30$ and σ is unknown proceeds as follows:

1. $H_0: \mu = \mu_0$.

2. H_a: Appropriate one- or two-tailed alternative.
3. Test statistic:

$$t = \frac{\bar{y} - \mu}{s/\sqrt{n}}.$$

4. Rejection region with $\alpha = P[\text{falsely rejecting } H_0]$:
    a. For $H_a: \mu > \mu_0$, reject H_0 if $t > t_\alpha$ based upon $n - 1$ degrees of freedom.
    b. For $H_a: \mu < \mu_0$, reject H_0 if $t < -t_\alpha$ based upon $n - 1$ degrees of freedom.
    c. For $H_a: \mu \neq \mu_0$, reject H_0 if $|t| > t_{\alpha/2}$ based upon $n - 1$ degrees of freedom.

Example 9.3
A new electronic device that requires two hours per item to produce on a production line has been developed by company A. While the new product is being run, profitable production time is used. Hence the manufacturer decides to produce only six new items for testing purposes. For each of the six items, the time to failure is measured, yielding the measurements 59.2, 68.3, 57.8, 56.5, 63.7, and 57.3 hours. Is there sufficient evidence to indicate that the new device has a mean life greater than 55 hours at the $\alpha = .05$ level?

Solution
To calculate the sample mean and standard deviation we need

$$\Sigma y_i = 362.8 \qquad \text{and} \qquad \Sigma y_i^2 = 22{,}043.60.$$

$$\bar{y} = \frac{1}{n} \Sigma y_i = \frac{362.8}{6} = \underline{\hspace{2cm}}. \qquad\qquad 60.47$$

$$s^2 = \frac{1}{n-1} \left[\Sigma y_i^2 - \frac{(\Sigma y_i)^2}{n} \right]$$

$$= \frac{1}{5} \left[22{,}043.60 - \frac{(362.8)^2}{6} \right]$$

$$= 21.2587$$

with $\qquad s = \sqrt{21.2587} = 4.61.$

The test proceeds as follows:

1. $H_0: \mu \underline{\hspace{2cm}}.$ $\qquad\qquad = 55$

2. $H_a: \mu \underline{\hspace{2cm}}.$ $\qquad\qquad > 55$

3. Test statistic:

$$t = \frac{\bar{y} - 55}{s/\sqrt{n}} \ .$$

4. Rejection region:

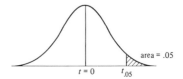

area = .05

$t = 0$ $t_{.05}$

2.015

Based on 5 degrees of freedom, reject H_0 if $t >$ _____.

Now calculate the value of the test statistic.

60.47; 5.47; 2.91
4.61

$$t = \frac{\bar{y} - 55}{s/\sqrt{n}} = \frac{-55}{\sqrt{6}} = \frac{-55}{1.88} = \underline{\qquad}.$$

is; reject
is

Since the observed value (is, is not) larger than 2.015, we (reject, do not reject) H_0. There (is, is not) sufficient evidence to indicate that the new device has a mean life greater than 55 hours at the 5% level of significance.

Confidence Interval for a Population Mean, μ
In estimating a population mean, one can use either a point estimator with bounds on error or an interval estimator having the required level of confidence.

  Small sample estimation of the mean of a *normal* population with σ *unknown* involves the statistic

$$\frac{\bar{y} - \mu}{s/\sqrt{n}}$$

$t; n - 1$

which has a _____ distribution with _____ degrees of freedom. The resulting $(1 - \alpha)100\%$ confidence interval estimator is given as

$$\left[\bar{y} \pm t_{\alpha/2} \frac{s}{\sqrt{n}} \right]$$

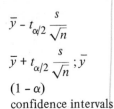

$\bar{y} - t_{\alpha/2} \dfrac{s}{\sqrt{n}}$

$\bar{y} + t_{\alpha/2} \dfrac{s}{\sqrt{n}} ; \bar{y}$

$(1 - \alpha)$
confidence intervals

where $t_{\alpha/2}$ is that value of t based upon $n - 1$ degrees of freedom having an area of $\alpha/2$ to its right. The lower confidence limit is _____ and the upper confidence limit is _____. The point estimator of μ is _____ and the bound on the error of estimation can be taken to be $t_{\alpha/2} s/\sqrt{n}$. A proper interpretation of a $(1 - \alpha)100\%$ confidence interval for μ would be stated as follows. In repeated sampling, (_____)100% of the _____ _____ so constructed would enclose the true value of the mean, μ.

Example 9.4
Using the data from Example 9.3 of this section, find a 95% confidence

interval estimate for μ, the mean life in hours for the new device.

Solution

The pertinent information from Example 9.3 follows.

$\bar{y} = 60.47$ d.f. = _____   5

$s/\sqrt{n} = 1.88$ $\alpha/2 = .025$

$t_{.025} = $ _____ .   2.571

The confidence interval will be found by using

$$\bar{y} \pm t_{.025} \, s/\sqrt{n} \, .$$

Substituting $\bar{y}$, $s/\sqrt{n}$, $t_{.025}$ we have

60.47 $\pm$ _____ (1.88)   2.571

60.47 $\pm$ _____   4.83

or

(_____ , _____).   55.64; 65.30

Self-Correcting Exercises 9A

1. In a random sample of ten cans of corn from supplier B, the average weight per can of corn was $\bar{y} = 9.4$ oz. with standard deviation, $s = 1.8$ oz. Does this sample contain sufficient evidence to indicate the mean weight is less than 10 oz. at the $\alpha = .01$ level? Find a 98% confidence interval for μ.
2. A school administrator claimed that the average time spent on a school bus by those students in his school district that rode school buses was 35 minutes. A random sample of twenty students who did ride the school buses yielded an average of 42 minutes riding time with a standard deviation of 6.2 minutes. Does this sample of size 20 contain sufficient evidence to indicate that the mean riding time is greater than 35 minutes at the 5% level of significance?
3. Find a 95% confidence interval for the mean time spent on a school bus using the data from Exercise 2.

9.4 Small Sample Inference Concerning the Difference Between Two Means, $\mu_1 - \mu_2$

Inferences concerning $\mu_1 - \mu_2$ based on small samples are founded upon the following assumptions:
1. Each population sampled has a _____ distribution.   normal
2. The population _____ are equal; that is, $\sigma_1^2 = \sigma_2^2$.   variances

$\bar{y}_1 - \bar{y}_2$

3. The samples are independently drawn.

An unbiased estimator for $\mu_1 - \mu_2$, regardless of sample size, is _____.
The standard deviation of this estimator is

$$\sqrt{\frac{\sigma_1^2}{n_1} + \frac{\sigma_2^2}{n_2}}.$$

When $\sigma_1^2 = \sigma_2^2$, we can replace σ_1^2 and σ_2^2 by a common variance, σ^2. Then the standard deviation of $\bar{y}_1 - \bar{y}_2$ becomes

σ

$$\sqrt{\frac{\sigma^2}{n_1} + \frac{\sigma^2}{n_2}} = (\underline{\hspace{1cm}}) \sqrt{\frac{1}{n_1} + \frac{1}{n_2}}.$$

If σ were known, then in testing an hypothesis concerning $\mu_1 - \mu_2$ we would use the statistic

$$z = \frac{(\bar{y}_1 - \bar{y}_2) - D_0}{\sigma \sqrt{\frac{1}{n_1} + \frac{1}{n_2}}}$$

where $D_0 = \mu_1 - \mu_2$. For small samples with σ unknown, we would use

$$t = \frac{(\bar{y}_1 - \bar{y}_2) - D_0}{s \sqrt{\frac{1}{n_1} + \frac{1}{n_2}}}$$

Student's t

σ^2

where s is the estimate of σ, calculated from the sample values. When the data are normally distributed, this statistic has a _____ _____ distribution with degrees of freedom the same as those available for estimating _____.

variance

variance
$s_1^2; s_2^2$
c

In selecting the best estimate (s^2) for σ^2 one has three immediate choices:
a. s_1^2, the sample _____ from population I.
b. s_2^2, the sample _____ from population II.
c. A combination of _____ and _____.
(a, b, c) is the best choice since it uses the information from both samples. A logical method of combining this information into one estimate, s^2, is

d. $$s^2 = \frac{(n_1 - 1)s_1^2 + (n_2 - 1)s_2^2}{(n_1 - 1) + (n_2 - 1)},$$

a weighted average of the sample variances using the degrees of freedom as weights.

The expression in d. can be written in another form by replacing s_1^2 and s_2^2 by their defining formulas. Then

$$s^2 = \frac{\sum\limits_{i=1}^{n_1} (y_i - \bar{y}_1)^2 + \sum\limits_{i=1}^{n_2} (y_i - \bar{y}_2)^2}{(n_1 - 1) + (n_2 - 1)}$$

In this form we see that we have pooled or added the sums of squared deviations from each sample and divided by the pooled degrees of freedom, $n_1 + n_2 - 2$. Hence s^2 is a pooled estimate of the common variance σ^2 and is based upon _____ degrees of freedom. Since our samples were drawn from normal populations, the statistic

$$n_1 + n_2 - 2$$

$$t = \frac{(\bar{y}_1 - \bar{y}_2) - (\mu_1 - \mu_2)}{s \sqrt{\dfrac{1}{n_1} + \dfrac{1}{n_2}}}$$

has a _____ _____ distribution with _____ degrees of freedom.

$$\text{Student's } t; n_1 + n_2 - 2$$

Example 9.5

A medical student conducted a diet study using two groups of 12 rats each as subjects. Group I received diet I while group II received diet II. After 5 weeks the student calculated the gain in weight for each rat. The data yielded the following information:

Group I	Group II
$\bar{y}_1$ = 6.8 oz.	$\bar{y}_2$ = 5.3 oz.
s_1 = 1.5 oz.	s_2 = 0.9 oz.
n_1 = 12	n_2 = 12

Do these data present sufficient evidence to indicate at the $\alpha = .05$ level that rats on diet I will gain more weight than those on diet II? Find a 90% confidence interval for $\mu_1 - \mu_2$, the mean difference in weight gained.

Solution

We shall take the gains in weight to be normally distributed with equal variances and calculate a pooled estimate for σ^2.

$$s^2 = \frac{(n_1 - 1) s_1^2 + (n_2 - 1) s_2^2}{n_1 + n_2 - 2}$$

$$= \frac{11(1.5)^2 + 11 (0.9)^2}{12 + 12 - 2} = \frac{33.66}{22} = \underline{\hspace{2cm}}$$

1.530

so that

$$s = \sqrt{1.530} = \underline{\hspace{2cm}} .$$

1.237

The test is as follows:

0

0

22
1.717

1.5; 0

1.5

2.97

reject H_0

1.717

1.5; .505

1.5; .867

.63; 2.37

1. $H_0: \mu_1 - \mu_2 = $ _____.

2. $H_a: \mu_1 - \mu_2 > $ _____.

3. Test statistic:

$$t = \frac{(\bar{y}_1 - \bar{y}_2) - D_0}{s\sqrt{\dfrac{1}{n_1} + \dfrac{1}{n_2}}}.$$

4. Rejection region:

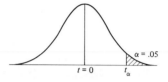

With $n_1 + n_2 - 2 = $ _____ degrees of freedom, we would reject H_0 if $t > $ _____. Calculate the test statistic.

$$t = \frac{(\bar{y}_1 - \bar{y}_2) - D_0}{s\sqrt{\dfrac{1}{n_1} + \dfrac{1}{n_2}}}$$

$$= \frac{(\text{\_\_\_\_\_}) - (\text{\_\_\_\_\_})}{1.237\sqrt{.1667}}$$

$$= \frac{(\text{\_\_\_\_\_})}{.505}$$

$$= \text{\_\_\_\_\_}.$$

Decision: _____.

To find a 90% confidence interval for $\mu_1 - \mu_2$, we need $t_{.05}$ based upon 22 degrees of freedom. $t_{.05} = $ _____. Hence we would use

$$(\bar{y}_1 - \bar{y}_2) \pm 1.717\, s\sqrt{\dfrac{1}{n1} + \dfrac{1}{n_2}}$$

$$\text{\_\_\_\_\_} \pm 1.717\,(\text{\_\_\_\_\_})$$

$$(\text{\_\_\_\_\_}) \pm (\text{\_\_\_\_\_}).$$

Therefore, a 90% confidence interval for $\mu_1 - \mu_2$ would be

$$(\text{\_\_\_\_\_}, \text{\_\_\_\_\_}).$$

Self-Correcting Exercises 9B

1. What are the assumptions required for the proper use of the statistic

$$t = \frac{(\bar{y}_1 - \bar{y}_2) - (\mu_1 - \mu_2)}{s\sqrt{\dfrac{1}{n_1} + \dfrac{1}{n_2}}} ?$$

2. In the process of making a decision to either continue operating or close a civic health center, a random sample of 25 people who had visited the center at least once was chosen and each person asked whether he felt the center should be closed. In addition, the distance between each person's place of residence and the health center was computed and recorded. Of the 25 people responding, 16 were in favor of continued operation. For these 16 people, the average distance from the center was 5.2 miles with a standard deviation of 2.8 miles. The remaining 9 people who were in favor of closing the center lived at an average of 8.7 miles from the center with a standard deviation of 5.3 miles. Do these data indicate that there is a significant difference in mean distance to the health center for these two groups?

3. Estimate the difference in mean distance to the health center for the two groups in Exercise 2 with a 95% confidence interval.

4. In investigating which of two presentations of subject matter to use in a computer programmed course, an experimenter randomly chose two groups of 18 students each, and assigned one group to receive presentation I and the second to receive presentation II. A short quiz on the presentation was given to each group and their grades recorded. Do the following data indicate that a difference in the mean quiz scores (hence, a difference in effectiveness of presentation) exists for the two methods?

	$\bar{y}$	s^2
Presentation I	81.7	23.2
Presentation II	77.2	19.8

9.5 A Paired-Difference Test

In many situations an experiment is designed so that a comparison of the effects of two "treatments" is made on the same person, on twin offspring, two animals from the same litter, two pieces of fabric from the same loom, or two plants of the same species grown on adjacent plots. Such experiments are designed so that the pairs of experimental units (people, animals, fabrics, plants) are as much alike as possible. By taking measurements on the two treatments within the relatively homogeneous pairs of experimental units, the difference in the measurements for the two treatments in a pair will primarily reflect the difference between _____ means rather than the differ- | treatment
ence between experimental units. This experimental design reduces the error of comparison and increases the quantity of information in the experiment.

reducing

To analyze such an experiment using the techniques of the last section would be incorrect. In planning this type of experiment, we *intentionally violate* the assumption that the measurements are *independent,* and hope that this violation will work to our advantage by (increasing, reducing) the variability of the differences of the paired observations. Consider the situation in which two sets of identical twin calves are selected for a diet experiment, where one of each set of twins is randomly chosen to be fed diet A while the other is given diet B. At the end of a given period of time, the calves are weighed and the data is presented for analysis.

	Diet		
Set	A	B	Difference
1	A_1	B_1	$A_1 - B_1$
2	A_2	B_2	$A_2 - B_2$

Now A_1 and B_1 are *not* independent since the calves are identical twins and as such have the same growth trend, weight-gain trend, and so on. Although A_1 could be larger or smaller than B_1, if A_1 were a large amount we would also expect B_1 to be a large amount. A_2 and B_2 are not independent for the same reason. However, by looking at the differences, $(A_1 - B_1)$ and $(A_2 - B_2)$, the characteristics of the twin calves no longer cloud the issue, since these differences would represent the difference due to the effects of the two treatments.

In using a paired-difference design we analyze the differences of the paired measurements and, in so doing, attempt to *reduce* the *variability* that would be present in two *randomly* selected groups without pairing. A test of the hypothesis that the difference in two population means, $\mu_1 - \mu_2$, is equal to a constant, D_0, is equivalent to a test of the hypothesis that the mean of the differences, μ_d, is equal to a constant, D_0. That is, $H_0: \mu_1 - \mu_2 = D_0$ is equivalent to $H_0: \mu_d = D_0$.

Example 9.6
In order to test the results of a conventional vs. new approach to the teaching of reading, twelve pupils were selected and matched according to I.Q., age, present reading ability, and so on. One from each of the pairs was assigned to the conventional reading program and the other to the new reading program. At the end of six weeks, their progress was measured by a reading test. Do the following data present sufficient evidence to indicate that the new approach is better than the conventional approach at the $\alpha = .10$ level?

Pair	Conventional	New	$d_i = N - C$
1	78	83	5
2	65	69	4
3	88	87	-1
4	91	93	2
5	72	78	6
6	59	59	0

Find a 95% confidence interval for the difference in mean reading scores.

Solution

We analyze the set of six differences as we would a single set of six measurements. The change in notation required is straightforward.

$$\sum_{i=1}^{6} d_i = 16 \qquad \sum_{i=1}^{6} d_i^2 = 82.$$

The sample mean is

$$\bar{d} = \frac{1}{6} \sum_{i=1}^{6} d_i = \underline{\hspace{2cm}}.$$

2.67

The sample variance of the differences is

$$s_d^2 = \frac{\Sigma d_i^2 - (\Sigma d_i)2/6}{5}$$

$$= \frac{82 - (16)^2/6}{5} = \frac{39.3333}{5} = \underline{\hspace{2cm}}$$

7.8667

$$s_d = \sqrt{7.8667} = \underline{\hspace{2cm}}.$$

2.80

The test is conducted as follows. Remember that $\mu_d = \mu_N - \mu_C$.

$$H_0: \mu_d = 0.$$

$$H_a: \mu_d \underline{\hspace{2cm}}.$$

> 0

Test statistic:

$$t = \frac{\bar{d} - 0}{s_d/\sqrt{n}}.$$

Rejection region: Based upon 5 degrees of freedom we will reject H_0 if the observed value of t is greater than $t_{.05} = \underline{\hspace{2cm}}$.

2.015

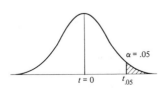

The sample value of t is

$$t = \frac{\bar{d} - 0}{s_d/\sqrt{n}}$$

2.67; 2.34

$$= \frac{\underline{\quad} - 0}{2.80/\sqrt{6}} = \frac{2.67}{1.14} = \underline{\quad\quad}.$$

reject

Since the value of the test statistic is greater than 2.015, we (reject, do not reject) H_0. This sample indicates that the new method appears to be superior to the conventional method at the $\alpha = .05$ level, if we assume that the reading test is a valid criterion upon which to base our judgment.

A 95% confidence interval for μ_d is estimated by

$$\bar{d} \pm t_{.025} \frac{s_d}{\sqrt{n}} .$$

2.571

Using sample values and $t_{.025} = \underline{\quad\quad}$ we have

2.571

$$2.67 \pm \underline{\quad\quad} (1.14)$$

2.93

$$2.67 \pm \underline{\quad\quad}.$$

-0.26
5.60

With 95% confidence, we estimate that μ_d lies within the interval $\underline{\quad\quad}$ to $\underline{\quad\quad}$.

Notice that in using a paired-difference analysis, the degrees of freedom for the critical value of t drop from $2n - 2$ for an unpaired design to $n - 1$ for the

$n - 1$
larger

paired, a loss of $(2n - 2) - (n - 1) = \underline{\quad\quad}$ degrees of freedom. This results in a (larger, smaller) critical value of t. Therefore, a larger value of the test statistic is needed to reject H_0. Fortunately, *proper* pairing will reduce $\sigma_{\bar{d}}$. Hence the paired-difference experiment results in both a loss and a gain of information. However, the *loss* of $(n - 1)$ degrees of freedom is usually far overshadowed by the gain in information when $\sigma_{\bar{d}}$ is substantially reduced.

Self-Correcting Exercises 9C

1. The owner of a small manufacturing plant is considering a change in salary base by replacing an hourly wage structure with a per-unit rate. He hopes that such a change will increase the output per worker, but has reservations about a possible decrease in quality under the per-unit plan. Before arriving at any decision, he forms 10 pairs of workers so that within each pair, the two workers have produced about the same number of items per day and their work is of comparable quality. From each pair, one worker is randomly selected to be paid as usual and the other to be paid on a per-unit basis. In addition to the number of items produced, a cumulative quality score for the items produced is kept for each worker. The quality scores follow. (A high score is indicative of high quality.)

	Rate	
Pair	Per Unit	Hourly
1	86	91
2	75	77
3	87	83
4	81	84
5	65	68
6	77	76
7	88	89
8	91	91
9	68	73
10	79	78

Do these data indicate that the average quality for the per-unit production is significantly lower than that based upon an hourly wage?

2. Refer to Exercise 1. The following data represent the average number of items produced per worker, based upon one week's production records:

	Rate	
Pair	Per Unit	Hourly
1	35.8	31.2
2	29.4	27.6
3	31.2	32.2
4	28.6	26.4
5	30.0	29.0
6	32.6	31.4
7	36.8	34.2
8	34.4	31.6
9	29.6	27.6
10	32.8	29.8

a. Estimate the mean difference in average daily output for the two pay scales with a 95% confidence interval.

b. Test the hypothesis that a per-unit pay scale increases production at the .05 level of significance.

9.6 Inference Concerning a Population Variance

In many cases, the measure of variability is more important than that of central tendency. For example, an educational test consisting of 100 items has a mean score of 75 with standard deviation of 2.5. $\mu = 75$ may sound impressive, but $\sigma = 2.5$ would imply that this test has very poor discriminating ability since approximately 95% of the scores would be between 70 and 80. In like manner a production line producing bearings with $\mu = .25$ inches and $\sigma = .5$ inches would produce many defective items; the fact that the bearings have a mean diameter of .25 inches would be of little value when the bearings are fitted together. *The precision of an instrument, whether it be an educational test or a machine, is measured by the standard deviation of the error of measurement.* Hence we proceed to a test of a population variance, σ^2.

unbiased

σ^2

nonsymmetric

zero

n

σ^2

The sample variance, s^2, is an _____ estimator for σ^2. To use s^2 for inference making, we find that in repeated sampling, the distribution of s^2 has the following properties:

1. $E(s^2) = $ _____.
2. The distribution of s^2 is (symmetric, nonsymmetric).
3. s^2 can assume any value greater than or equal to _____.
4. The shape of the distribution changes for different values of _____ and _____.
5. In sampling from a *normal* population, s^2 is independent of the population mean, μ, and the sample mean, $\bar{y}$. As with the z-statistic, the distribution for s^2 when sampling from a normal population can be standardized by using

$$\chi^2 = \frac{(n-1)s^2}{\sigma^2} \, ,$$

the chi-square variable having the following properties in repeated sampling.

$n-1$

1. $E(\chi^2) = d.f. = $ _____.

nonsymmetric

2. The distribution of χ^2 is (symmetric, nonsymmetric).

0

3. $\chi^2 \geqslant$ _____.

4. The distribution of χ^2 depends upon the degrees of freedom, $n - 1$. Since χ^2 does not have a symmetric distribution, critical values of χ^2 have been tabulated for both the upper and lower tails of the distribution in Table 5 of your text. The degrees of freedom are listed along both the right and left margins of the table. Across the top margin are values, χ_α^2, indicating a value of χ^2 having an area equal to α to its right, that is,

$$P[\chi^2 > \chi_\alpha^2] = \alpha.$$

Example 9.7
Use Table 5 to find the following critical values of χ^2:

		α	d.f.	χ_α^2
5.99147	a.	.05	2	_____
2.55821	b.	.99	10	_____
37.5662	c.	.01	20	_____
18.4926	d.	.95	30	_____
1.734926	e.	.995	9	_____
27.4884	f.	.025	15	_____
45.5585	g.	.005	24	_____
10.0852	h.	.90	17	_____

The statistical test of an hypothesis concerning a population variance σ^2 at the α level of significance is given as follows:

1. $H_0: \sigma^2 = \sigma_0^2.$

2. H_a: Appropriate one- or two-tailed test.
3. Test statistic:

$$\chi^2 = \frac{(n-1)s^2}{\sigma_0^2}$$

4. Rejection region:
   a. For H_a: $\sigma^2 > \sigma_0^2$, reject H_0 if $\chi^2 > \chi_\alpha^2$ based on $n-1$ degrees of freedom.
   b. For H_a: $\sigma^2 < \sigma_0^2$, reject H_0 if $\chi^2 < \chi_{(1-\alpha)}^2$ based on $n-1$ degrees of freedom.
   c. For H_a: $\sigma^2 \neq \sigma_0^2$, reject H_0 if $\chi^2 > \chi_{\alpha/2}^2$ or $\chi^2 < \chi_{(1-\alpha/2)}^2$ based on $n-1$ degrees of freedom.

Example 9.8
A producer of machine parts claimed that the diameters of the connector rods produced by his plant had a variance of at most .03 in.^{2} A random sample of 15 connector rods from his plant produced a sample mean and variance of 0.55 in. and .053 in.^{2}, respectively. Is there sufficient evidence to reject his claim at the $\alpha = .05$ level of significance?

Solution
1. Collecting pertinent information:

$$s^2 = .053 \text{ in.}^2$$

$$d.f. = n - 1 = 14.$$

$$\sigma_0^2 = .03 \text{ in.}^2$$

2. The test of the hypothesis is given as:

$$H_0: \sigma^2 = .03.$$

$$H_a: \sigma^2 \underline{\hspace{2cm}} .03.$$ $>$

   Test statistic:

$$\chi^2 = \frac{(n-1)s^2}{\sigma_0^2} \, .$$

   Rejection region:

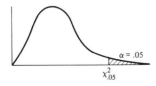
$\alpha = .05$
$\chi_{.05}^2$

For 14 degrees of freedom, we shall reject H_0 if $\chi^2 \geqslant \underline{\hspace{2cm}}$.   23.6848

Calculate:

$$\chi^2 = \frac{(n-1)s^2}{\sigma_0^2}$$

.053

.03

$$= \frac{14(\underline{\hspace{2cm}})}{(\underline{\hspace{2cm}})}$$

$$= 24.733.$$

Reject

Decision: (Reject, Do not reject) H_0 since

$$24.733 > \chi^2_{.05} = 23.6848.$$

is

The data produced sufficient evidence to reject H_0. Therefore, we can conclude that the variance of the rod diameters (is, is not) greater than .03 in.^{2}.

The sample variance s^2 is an unbiased point estimator for the population variance σ^2. Utilizing the fact that $(n-1)s^2/\sigma^2$ has a chi-square distribution with $(n-1)$ degrees of freedom, we can show that a $(1-\alpha)100\%$ confidence interval for σ^2 is

$$\frac{(n-1)s^2}{\chi^2_U} < \sigma^2 < \frac{(n-1)s^2}{\chi^2_L},$$

$n-1$

where χ^2_U is the tabulated value of the chi-square random variable based upon _____ degrees of freedom having an area equal to $\alpha/2$ to its right, while χ^2_L is the tabulated value from the same distribution having an area of $\alpha/2$ to its left or equivalently, an area of $1 - \alpha/2$ to its right.

Example 9.9
Find a 95% confidence interval estimate for the variance of the rod diameters from Example 9.8.

Solution
From Example 9.8 the estimate of σ^2 was $s^2 = .053$ with 14 degrees of freedom. For a confidence coefficient of .95 we need

5.62872

$$\chi^2_L = \chi^2_{.975} = \underline{\hspace{2cm}}$$

and

26.1190

$$\chi^2_U = \chi^2_{.025} = \underline{\hspace{2cm}}.$$

1. Using the confidence interval estimator

$$\frac{(n-1)s^2}{\chi^2_U} < \sigma^2 < \frac{(n-1)s^2}{\chi^2_L}$$

we have

$$\frac{14\,(.053)}{(\underline{\quad\quad})} < \sigma^2 < \frac{14\,(.053)}{(\underline{\quad\quad})}$$

26.1190; 5.62872

$$\underline{\quad\quad} < \sigma^2 < \underline{\quad\quad}\,.$$

.028; .132

2. By taking square roots of the upper and lower confidence limits, we have an equivalent confidence interval for the standard deviation σ. For this problem

$$\underline{\quad\quad} < \sigma < \underline{\quad\quad}\,.$$

.167; 363

Comment. Although the sample variance is an unbiased point estimator for σ^2, notice that the confidence interval estimator for σ^2 *is not symmetrically located about* $\hat{\sigma}^2$ as was the case with confidence intervals which were based upon the z or t distributions. This follows from the fact that a chi-square distribution is not symmetric while the z and t distributions are symmetric.

Self-Correcting Exercises 9D

1. In an attempt to assess the variability in the time until a pain reliever became effective for a patient, a doctor, on five different occasions, pre-scribed a controlled dosage of the drug for his patient. The five measure-ments recorded for the time until effective relief were: 20.2, 15.7, 19.8, 19.2, 22.7 minutes. Would these measurements indicate that the standard deviation of the time until effective relief was less than 3 minutes?

2. An educational testing service, in developing a standardized test, would like the test to have a standard deviation of at least 10. The present form of the test has produced a standard deviation of $s = 8.9$ based upon $n = 30$ test scores. Should the present form of the test be revised based upon these sample data?

3. A quick technique for determining the concentration of a chemical solu-tion has been proposed to replace the standard technique which takes much longer. In testing a standardized solution, 30 determinations using the new technique produced a standard deviation of $s = 7.3$ parts per million.
   a. Does it appear that the new technique is less sensitive (has larger vari-ability) than the standard technique whose standard deviation is $\sigma = 5$ ppm?
   b. Estimate the true standard deviation for the new technique with a 95% confidence interval.

9.7 Comparing Two Population Variances

An experimenter may wish to compare the variability of two testing pro-cedures or compare the precision of one manufacturing process with another. One may also wish to compare two population variances prior to using a t-test.

To test the hypothesis of equality of two population variances,

$$H_0: \sigma_1^2 = \sigma_2^2,$$

we need the following assumptions:

normal

independent

1. Each population sampled has a _____ distribution.
2. The samples are _____.

The statistic, s_1^2/s_2^2, is used to test

$$H_0: \sigma_1^2 = \sigma_2^2.$$

large

small

A _____ value of this statistic implies that $\sigma_1^2 > \sigma_2^2$; a _____ value of this statistic implies that $\sigma_1^2 < \sigma_2^2$; while a value of the statistic close to one (1) implies that $\sigma_1^2 = \sigma_2^2$. In repeated sampling this statistic has an F distribution when $\sigma_1^2 = \sigma_2^2$ with the following properties:

nonsymmetric

$s_1^2; s_2^2$

zero

1. The distribution of F is (symmetric, nonsymmetric).
2. The shape of the distribution depends upon the degrees of freedom associated with _____ and _____.
3. F is always greater than or equal to _____.

The tabulation of critical values of F is complicated by the fact that the distribution is nonsymmetric and must be indexed according to the values of v_1 and v_2, the degrees of freedom associated with the numerator and denominator of the F-statistic. As we shall see, however, it will be sufficient to have only right-tailed critical values of F for the various combinations of v_1 and v_2. Tables 6 and 7 have tabulated right-tailed critical values, F_α, for the F-statistic where F_α is that value of F having an area equal to α to its right, based on v_1 and v_2, the degrees of freedom associated with the *numerator* and *denominator* of F, respectively. F_α satisfies the relationship $P[F > F_\alpha] = \alpha$.

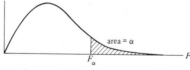

area = α

F_α

Table 6 has values of F_α for $\alpha = .05$, and various values of v_1 and v_2 between 1 and ∞, while Table 7 has the same information for $\alpha = .01$.

Example 9.10

Find the value of F based upon $v_1 = 5$ and $v_2 = 7$ degrees of freedom such that

$$P[F > F_{.05}] = .05.$$

Solution

1. We wish to find a critical value of F with an area $\alpha = .05$ to its right based on $v_1 = 5, v_2 = 7$ degrees of freedom. Therefore, we will use Table 6.
2. Values of v_1 are found along the *top* margin of the table while values of v_2 appear on both the right *and* left margins of the table. Find the value of $v_1 = 5$ along the top margin and cross-index this value with $v_2 = 7$ along the left margin to find $F_{.05(5,7)} = 3.97$.

Example 9.11

Find the critical right-tailed values of F for the following:

	v_1	v_2	α	F_α	
a.	5	2	.05	_____	19.30
b.	7	15	.05	_____	2.71
c.	20	10	.01	_____	4.41
d.	30	40	.05	_____	1.74
e.	17	13	.01	_____	3.76

We can always avoid using left-tailed critical values of the F distribution by using the following approach. In testing $H_0: \sigma_1^2 = \sigma_2^2$ against the alternative $H_a: \sigma_1^2 > \sigma_2^2$, we would reject H_0 only if s_1^2/s_2^2 is *too large* (larger than a right-tailed critical value of F). In testing $H_0: \sigma_1^2 = \sigma_2^2$ against $H_a: \sigma_1^2 < \sigma_2^2$, we would reject H_0 only if s_2^2/s_1^2 were *too large*. In testing $H_0: \sigma_1^2 = \sigma_2^2$ against the two-tailed alternative, $H_a: \sigma_1^2 \neq \sigma_2^2$, we shall agree to *designate the population which produced the larger sample variance as population* 1 *and the larger sample variance as* s_1^2. We then agree to reject H_0 if s_1^2/s_2^2 is *too large.*

When we agree to designate the population with the larger sample variance as population 1, the test of $H_0: \sigma_1^2 = \sigma_2^2$ versus $H_a: \sigma_1^2 \neq \sigma_2^2$ using s_1^2/s_2^2 will be right-tailed. However, in so doing we must remember that the tabulated tail area must be doubled to get the actual significance level of the test. For example, if the critical right-tailed value of F has been found from Table 6, the actual significance level of the test will be $\alpha = 2(.05) = $ _____ . If the critical value comes from Table 7, the actual level will be $\alpha = 2(.01) = $ _____ .

.10

.02

Example 9.12

An experimenter has performed a laboratory experiment using two groups of rats. One group was given a standard treatment while the second received a newly developed treatment. Wishing to test the hypothesis $H_0: \mu_1 = \mu_2$, the experimenter suspects that the population variances are not equal, an assumption necessary for using the t-statistic in testing the equality of the means. Use the following data to test if the experimenter's suspicion is warranted at the $\alpha = .02$ level:

Old Treatment	New Treatment
$s = 2.3$	$s = 5.8$
$n = 10$	$n = 10$

Solution

This problem involves a test of the equality of two population variances. Let population 1 be the population receiving the new treatment.

1. $H_0: \sigma_1^2 = \sigma_2^2$.

2. $H_a: \sigma_1^2 \neq \sigma_2^2$.

3. Test statistic:

$$F = s_1^2/s_2^2.$$

4. Rejection region: With $v_1 = v_2 = 9$ degrees of freedom, reject H_0 if

5.35 $\qquad\qquad F > F_{.01} = \underline{\hspace{2cm}}.$

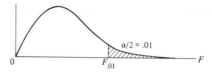

Calculate the test statistic.

$$F = s_1^2/s_2^2$$

$$= (5.8)^2/(2.3)^2$$

33.64; 5.29 $\qquad\qquad = (\underline{\hspace{2cm}}) / (\underline{\hspace{2cm}})$

6.36 $\qquad\qquad = \underline{\hspace{2cm}}.$

greater; is $\qquad\qquad$ Decision: Since F is (greater, less) than $F_{.01} = 5.35$, $H_0: \sigma_1^2 = \sigma_2^2$ (is, is not) rejected.

Since the population variances were judged to be different the experimenter is not justified in using the t-statistic to test $H_0: \mu_1 - \mu_2 = 0$. He must resort to other methods, several of which will be discussed in Chapter 14.

Utilizing the fact that $\dfrac{s_1^2}{s_2^2} \cdot \dfrac{\sigma_2^2}{\sigma_1^2}$ has an F distribution with $v_1 = (n_1 - 1)$ and $v_2 = (n_2 - 1)$ degrees of freedom, it can be shown that a $(1 - \alpha)100\%$ confidence interval for $\dfrac{\sigma_1^2}{\sigma_2^2}$ is

$$\frac{s_1^2}{s_2^2} \frac{1}{F_{v_1 v_2}} < \frac{\sigma_1^2}{\sigma_2^2} < \frac{s_1^2}{s_2^2} F_{v_2 v_1}$$

where $F_{v_1 v_2}$ is the tabulated value of F with v_1 and v_2 degrees of freedom and area $\alpha/2$ to its right, and $F_{v_2 v_1}$ is the tabulated value of F with v_2 and v_1 degrees of freedom having area $\alpha/2$ to its right.

Example 9.13

A comparison of the precisions of two machines developed for extracting juice from oranges is to be made using the following data:

Machine A	Machine B
$s^2 = 3.1$ oz.^{2}	$s^2 = 1.4$ oz.^{2}
$n = 25$	$n = 25$

Is there sufficient evidence to indicate that $\sigma_A^2 > \sigma_B^2$ at the $\alpha = .05$ level? Find a 90% confidence interval for σ_A^2/σ_B^2.

Solution

Let population 1 be the population of measurements on machine A. The test would proceed as follows:

$$H_0: \sigma_1^2 = \sigma_2^2.$$

$$H_a: \sigma_1^2 \underline{\qquad} \sigma_2^2.$$

$>$

Test statistic:

$$F = s_1^2/s_2^2.$$

Rejection region: Based upon $v_1 = v_2 = \underline{\qquad}$ degrees of freedom, we shall reject H_0 if $F > F_{.05}$ with $F_{.05} = \underline{\qquad}$.

24

1.98

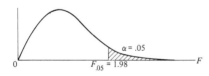

$\alpha = .05$

$F_{.05} = 1.98$

The value of the statistic is

$$F = \frac{s_1^2}{s_2^2} = \frac{3.1}{1.4} = \underline{\qquad}.$$

2.21

Decision: We reject H_0 and conclude that the variability of machine A (is, is not) greater than that of machine B.

is

A 90% confidence interval for $\sigma_A^2/\sigma_B^2 = \sigma_1^2/\sigma_2^2$ is

$$\frac{s_1^2}{s_2^2} \cdot \frac{1}{F_{24,24}} < \frac{\sigma_A^2}{\sigma_B^2} < \frac{s_1^2}{s_2^2} F_{24,24}$$

$$\frac{2.21}{1.98} < \frac{\sigma_A^2}{\sigma_B^2} < 2.21\,(1.98)$$

$$\underline{\qquad} < \frac{\sigma_A^2}{\sigma_B^2} < \underline{\qquad}.$$

1.12; 4.38

Self-Correcting Exercises 9E

1. Refer to Exercise 2, Self-Correcting Exercises 9B. In using the t-statistic in testing an hypothesis concerning $\mu_1 - \mu_2$, one assumes that $\sigma_1^2 = \sigma_2^2$. Based upon the sample information, could you conclude that this assumption had been met for this problem?

2. An experiment to explore the pain thresholds to electrical shock for males and females resulted in the following data summary:

	Males	Females
n	10	13
$\bar{y}$	15.1	12.6
s^2	11.3	26.9

Do these data supply sufficient evidence to indicate a significant difference in variability of thresholds for these two groups at the 10% level of significance?

EXERCISES

1. Why can we say that the test statistics employed in Chapter 8 are approximately normally distributed?

2. What assumptions are made when Student's t-statistic is used to test an hypothesis concerning a population mean, μ?

3. How does one determine the degrees of freedom associated with a t-statistic?

4. Ten butterfat determinations for brand G milk were carried out yielding $\bar{y} = 3.7\%$ and $s = 1.7\%$. Do these results produce sufficient evidence to indicate that brand G milk contains on the average less than 4.0% butterfat? (Use $\alpha = .05$.)

5. Refer to Exercise 4. Estimate the mean percentage of butterfat for brand G milk with a 95% confidence interval.

6. An experimenter has developed a new fertilizing technique that should increase the production of cabbages. Do the following data produce sufficient evidence to indicate that the mean weight of those cabbages grown using the new technique is greater than the mean weight of those grown using the standard technique?

Population I (New Technique)	Population II (Standard Technique)
$n_1 = 16$	$n_2 = 10$
$\bar{y}_1 = 33.4$ oz.	$\bar{y}_2 = 31.8$ oz.
$s_1 = 3$ oz.	$s_2 = 4$ oz.

7. Find a 90% confidence interval for the difference in means, $\mu_1 - \mu_2$, for the data given in Exercise 6.

8. To test the comparative brightness of two red dyes, nine samples of cloth were taken from a production line and each sample was divided into two pieces. One of the two pieces in each sample was randomly chosen and red dye 1 applied; red dye 2 was applied to the remaining piece. The

following data represent a "brightness score" for each piece. Is there sufficient evidence to indicate a difference in mean brightness scores for the
two dyes?

Sample	Dye 1	Dye 2
1	10	8
2	12	11
3	9	10
4	8	6
5	15	12
6	12	13
7	9	9
8	10	8
9	15	13

9. To test the effect of alcohol in increasing the reaction time to respond to a
given stimulus, the reaction times of seven persons were measured. After
consuming three ounces of 40% alcohol, the reaction time for each of the
seven persons was measured again. Do the following data indicate that the
mean reaction time after consuming alcohol was greater than the mean
reaction time before consuming alcohol? (Use $\alpha = .05$.)

Person	Before (time in seconds)	After (time in seconds)
1	4	7
2	5	8
3	5	3
4	4	5
5	3	4
6	6	5
7	2	5

0. A manufacturer of odometers claimed that mileage measurements indicated on his instruments had a variance of at most .53 miles per ten
miles traveled. An experiment, consisting of eight runs over a measured
ten-mile stretch, was performed in order to check the manufacturer's
claim. The variance obtained for the eight runs was 0.62. Does this provide sufficient evidence to indicate that $\sigma^2 > .53$? (Use $\alpha = .05$.)

1. Construct a 99% confidence interval estimate for σ^2 in Exercise 10.

2. In a test of heat resistance involving two types of metal paint, two groups
of ten metal strips were randomly formed. Group one was painted with
type I paint, while group two was painted with type II paint. The metal
strips were placed in an oven in random order, heated, and the temperature
at which the paint began to crack and peel recorded for each strip. Do the
following data indicate that the variability in the critical temperatures
differs for the two types of paint?

	$\bar{y}$	s^2	n
Type I	280.1°F	93.2	10
Type II	269.9°F	51.9	10

13. Construct a 98% confidence interval for σ_1^2/σ_2^2 for the data given in Exercise 12.

14. In an attempt to reduce the variability of machine parts produced by process A, a manufacturer has introduced process B (a modification of A). Do the following data based on two samples of 25 items indicate that the manufacturer has achieved his goal?

	n	s^2
Process A	25	6.57
Process B	25	3.19

Chapter 10

LINEAR REGRESSION AND

CORRELATION

10.1 Introduction

We have investigated the problem of making inferences about population parameters in the case of large and small sample sizes. We will now consider another aspect of this problem. Suppose that the expected value of a random variable y, $E(y)$, depends upon the values assigned to other variables, $x_1, x_2, \ldots, x_k$. Then we say that a functional relationship exists between $E(y)$ and $x_1, x_2, \ldots, x_k$. Since the values of $E(y)$ depend upon the values assumed by $x_1, x_2, \ldots, x_k$, $E(y)$ is called the *dependent* variable and $x_1, x_2, \ldots, x_k$ are called the *independent* variables. We restrict our investigation to the case where $E(y)$ is a *linear* function of one variable, x. By linear, we mean that the relationship between $E(y)$ and x can be described by a straight line.

Review: The Algebraic Representation of a Straight Line
To understand the development of the following linear models, you must be familiar with the algebraic representation of a straight line and its properties.
    The mathematical equation for a straight line is

$$y = \beta_0 + \beta_1 x,$$

where x is the independent variable, y is the dependent variable, and β_0 and β_1 are fixed constants. When values of x are substituted into this equation, pairs of numbers, (x_i, y_i), are generated which, when plotted or graphed on a rectangular coordinate system, form a straight line.

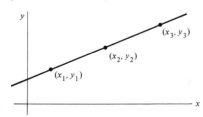

Consider the graph of a linear equation, $y = \beta_0 + \beta_1 x$, shown below.

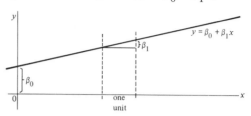

intercept

slope

1. By setting $x = 0$, we have $y = \beta_0 + \beta_1(0) = \beta_0$. Because the line intercepts or cuts the y-axis at the value $y = \beta_0$, β_0 is called the y-_____.
2. The constant, β_1, represents the increase in y for a one-unit increase in x and is called the _____ of the line.

Example 10.1
Plot the equation $y = 1 + 0.5x$ on a rectangular coordinate system.

Solution
Two points are needed to uniquely determine a straight line and therefore a minimum of two points must be found. A third point is usually found as a check on calculations.
1. Using 0, 2 and 4 as values of x, find the corresponding values of y.

1

2

3

When $x = 0$, $= 1 + 0.5(0) =$ _____.

When $x = 2$, $y = 1 + 0.5(2) =$ _____.

When $x = 4$, $y = 1 + 0.5(4) =$ _____.

2. Plot these points on a rectangular coordinate system and join them by using a straightedge.

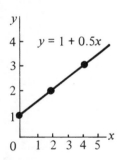

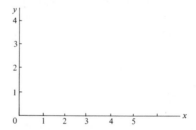

Practice plotting the following linear equations on a rectangular coordinate system:

a. $y = -1 + 3x$.

b. $y = 2 - x$.

c. $y = -0.5 - 0.5x$.

d. $y = x$.

e. $y = 0.5 + 2x$.

10.2 A Simple Linear Probabilistic Model

Suppose one is given a set consisting of n pairs of values for x and y, each pair representing the value of a response, y, for a given value of x. Plotting these points might result in the following scatter diagram:

Someone might say that these points appear to lie on a straight line. This person would be hypothesizing that a *model* for the relationship between x and y is of the form

$$y_i = \beta_0 + \beta_1 x_i \qquad i = 1, 2, \ldots, n.$$

According to this model, for a given value of x, the value of y is *uniquely* determined. Therefore this is called a _____ model.

Another person might say that these points appear to be *deviations* about a straight line, hypothesizing the model

$$y_i = \beta_0 + \beta_1 x_i + \epsilon_i \qquad i = 1, 2, \ldots, n,$$

where ϵ_i represents the deviation of the i-th point (x_i, y_i) from the straight line $y = \beta_0 + \beta_1 x$.

Suppose that we were able to make four observations on y at each of the values x_1, x_2, and x_3. We might observe the following twelve pairs of values:

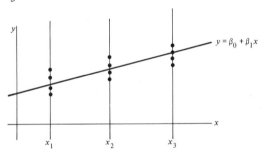

To account for what appear to be random deviations about the line $y = \beta_0 + \beta_1 x$, we shall consider the deviations to be *random errors* with the following properties:

1. For any fixed value of x, in repeated sampling, the random errors have a mean of zero and a variance equal to σ^2.
2. Any two random errors are independent in the probabilistic sense.
3. Regardless of the value of x, the random errors have the same *normal* distribution with mean zero and variance σ^2.

Since this model uses a random error component having a probability distribution, it is referred to as a _____ model.

The probabilistic model assumes that the average value of y is linearly related to x and the observed values of y will deviate above and below the line

$$E(y) = \beta_0 + \beta_1 x$$

by a random amount. The random components all have the same normal distribution and are independent of each other. According to the properties

deterministic

probabilistic

given above, repeated observations on y at the values x_1, x_2, and x_3 would result in the following visual representation of the random errors:

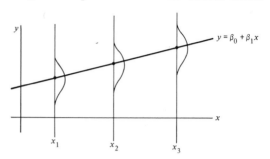

The probabilistic model appears to be the model which best describes the data and we now proceed to find an estimate for this prediction equation, the regression line,

$$\hat{y} = \hat{\beta}_0 + \hat{\beta}_1 x.$$

10.3 The Method of Least Squares

The criterion used for estimating β_0 and β_1 in the model

$$y_i = \beta_0 + \beta_1 x_i + \epsilon_i$$

is to find an estimated line

$$\hat{y}_i = \hat{\beta}_0 + \hat{\beta}_1 x_i$$

that in some sense minimizes the deviations of the observed values of y from the fitted line. If the deviation of the i-th observed value from the fitted value is $(y_i - \hat{y}_i)$, we define the best estimated line as one that minimizes the sum of squares of the deviations of the observed values of y from the fitted values of y. The quantity

$$\sum_{i=1}^{n} (y_i - \hat{y}_i)^2$$

represents the sum of squares of deviations of the observed values of y from the fitted values and is called the sum of squares for error (SSE).

$$SSE = \sum_{i=1}^{n} (y_i - \hat{y}_i)^2 = \sum_{i=1}^{n} [y_i - (\hat{\beta}_0 + \hat{\beta}_1 x_i)]^2$$

The values of $\hat{\beta}_0$ and $\hat{\beta}_1$ are determined mathematically so that SSE will be minimum.

This process of minimization is called the method of least squares and produces estimates of β_0 and β_1. If we agree that all summations will be with respect to i as the variable of summation, $i = 1, 2, \ldots, n$, then the least squares estimates of β_1 and β_0 are:

$$\hat{\beta}_1 = \frac{SS_{xy}}{SS_x} \qquad \text{and} \qquad \hat{\beta}_0 = \bar{y} - \hat{\beta}_1 \bar{x}$$

where

$$SS_{xy} = \Sigma(x_i - \bar{x})(y_i - \bar{y}) = \Sigma x_i y_i - \frac{(\Sigma x_i)(\Sigma y_i)}{n}$$

and

$$SS_x = \Sigma(x_i - \bar{x})^2 = \Sigma x_i^2 - \frac{(\Sigma x_i)^2}{n}.$$

Example 10.2

In this chapter we will use the following example to illustrate each type of problem encountered. Be ready to refer to the information tabulated on this page. For the following data, find the best fitting line, $\hat{y} = \hat{\beta}_0 + \hat{\beta}_1 x$:

x_i	y_i	x_i^2	y_i^2	$x_i y_i$	
2	1	4	1	2	
3	3	9	9	9	
5	4	25	16	20	
7	7	49	49	49	
9	10	81	100	90	
Σ _____ _____		168	175	170	26; 25
$\bar{x} =$ _____ $\bar{y} =$ _____					5.2; 5

Solution

1. First find all the sums needed in the computations.

$$SS_{xy} = \Sigma x_i y_i - \frac{(\Sigma x_i)(\Sigma y_i)}{n}$$

$$= 170 - \frac{(\underline{\hspace{1cm}})(\underline{\hspace{1cm}})}{5} \qquad \text{26; 25}$$

$$= 170 - \underline{\hspace{1cm}} \qquad \text{130}$$

$$= \underline{\hspace{1cm}} \qquad \text{40}$$

$$SS_{..} = \Sigma x_i^2 - \frac{(\Sigma x_i)^2}{n}$$

$$= \underline{\hspace{1cm}} - \frac{(26)^2}{5} \qquad \text{168}$$

$$= \underline{\hspace{1cm}} - 135.2 \qquad \text{168}$$

$$= \underline{\hspace{1cm}} \qquad \text{32.8}$$

2. $$\hat{\beta}_1 = \frac{SS_{xy}}{SS_x}$$

$$= \frac{4.0}{32.8}$$

$$= \underline{\hspace{1cm}} \qquad \text{1.22}$$

3. $\hat{\beta}_0 = \bar{y} - \hat{\beta}_1 \bar{x}$

5; 5.2   = (_____) − 1.22(_____)

5; 6.34   = (_____) − (_____)

           = −1.34.

4. The best fitting line is

−1.34 + 1.22x $\hat{y} =$ _____ .

We can now use the equation

$$\hat{y} = -1.34 + 1.22x$$

y
bounds; error

to predict values of _____ for values of x in the interval $2 \leqslant x \leqslant 9$. However, we also need to place _____ of _____ on this prediction. To do this we need σ^2, or its estimator, s^2.

Self-Correcting Exercises 10A

1. The registrar at a small university noted that the pre-enrollment figures and the actual enrollment figures for the past 6 years (in hundreds of students) were

x: Pre-enrollment	30	35	42	48	50	51
y: Actual enrollment	33	41	46	52	59	55

    a. Plot these data. Does it appear that a linear relationship exists between x and y?
    b. Find the least-squares line, $\hat{y} = \hat{\beta}_0 + \hat{\beta}_1 x$.
    c. Using the least-squares line, predict the actual number of students enrolled if the pre-enrollment figure is 5,000 students.

2. An entomologist, interested in predicting cotton harvest using the number of cotton bolls per quadrate counted during the middle of the growing season, collected the following data, where y is the yield in bales of cotton per field quadrate and x is hundreds of cotton bolls per quadrate counted during mid-season:

y	21	17	20	19	15	23	20
x	5.5	2.8	4.7	4.3	3.7	6.1	4.5

    a. Fit the least-squares line $\hat{y} = \hat{\beta}_0 + \hat{\beta}_1 x$ using these data.
    b. Plot the least-squares line and the actual data on the same graph. Comment on the adequacy of the least-squares predictor to describe these data.

3. Refer to Exercise 2. The same entomologist also had available a measure of

the number of damaging insects present per quadrate during a critical time in the development of the cotton plants. The data follow.

y: yield	21	17	20	19	15	23	20
x: insects	11	20	13	12	18	10	12

a. Fit the least-squares line to these data.
b. Plot the least-squares line and the actual data points on the same graph. Does it appear that the prediction line adequately describes the relationship between yield (y) and the number of insects present (x)?

10.4 Calculating s^2, the Estimator of σ^2.

To estimate σ^2, we use SSE, the sum of squared deviations about the line $\hat{y} = \hat{\beta}_0 + \hat{\beta}_1 x$. The n pairs of data points provide n degrees of freedom for estimation. Having estimated β_0 and β_1, we now have _____ degrees of freedom to estimate σ^2. Therefore, the estimator σ^2 is

$n - 2$

$$s^2 = \frac{SSE}{n-2} = \frac{\Sigma(y_i - \hat{y}_i)^2}{n-2}.$$

The computational form for the quantity SSE is

$$SSE = SS_y - \hat{\beta}_1 SS_{xy}$$

where

$$SS_y = \Sigma(y_i - \bar{y})^2 = \Sigma y_i^2 - \frac{(\Sigma y_i)^2}{n}.$$

SS_{xy} is the numerator used in computing $\hat{\beta}_1$ and has already been found.

Example 10.3
Calculate s^2 for our data.

Solution
1. First calculate the quantity SSE using the computational form.

$$SSE = SS_y - \hat{\beta}_1 SS_{xy}$$

$$= \left[175 - \frac{(\underline{\quad})^2}{5}\right] - 1.22\,[\underline{\quad}]$$ 25; 400

$$= [\underline{\quad}] - [\underline{\quad}]$$ 50; 48.8

$$= \underline{\quad}$$ 1.2

2. $$s^2 = \frac{SSE}{d.f.} = \frac{SSE}{n - 2}$$

1.2; 3

$$s^2 = \frac{()}{()} = .4 \quad \text{and} \quad s = \sqrt{.4} = .63.$$

10.5 Inferences Concerning the Slope of the Line, β_1

y

x

The slope, β_1, is the average increase in _____ for a one-unit increase in _____. The question of the existence of a linear relationship between x and y must be phrased in terms of the slope β_1. If no linear relationship exists between x and y, then $\beta_1 = 0$. Hence a test of the existence of a *linear*

0

0

relationship between x and y is given as $H_0: \beta_1 = $ _____ versus $H_a:$ $H_a: \beta_1 \neq$ _____.

When the random error ϵ is *normally* distributed, the estimator $\hat{\beta}_1$ has the following properties:

normal

$\beta_1; \beta_1$

1. $\hat{\beta}_1$ has a _____ distribution.
2. $\hat{\beta}_1$ is an unbiased estimator for _____ so that $E(\hat{\beta}_1) = $ _____.
3. The variance of $\hat{\beta}_1$ is

$$\sigma_{\hat{\beta}_1}^2 = \frac{\sigma^2}{SS_x}.$$

The following test statistics can be constructed using the fact that $\hat{\beta}_1$ is a *normally* distributed, *unbiased* estimator of β_1.

1. $$z = \frac{\hat{\beta}_1 - \beta_1}{\sigma/\sqrt{SS_x}} \quad \text{if } \sigma^2 \text{ is known.}$$

2. $$t = \frac{\hat{\beta}_1 - \beta_1}{s/\sqrt{SS_x}} \quad \text{if } s^2 \text{ is used to estimate } \sigma^2, \text{ and hence to estimate } \sigma_{\hat{\beta}_1}^2.$$

Since σ^2 is rarely known, we can test the hypothesis for linearity at the α level using the statistic given in 2., which has a Student's t distribution with

$n - 2$

_____ degrees of freedom. A test of the hypothesis $H_0: \beta_1 = 0$ versus $H_a: \beta_1 \neq 0$ is given as:

1. $H_0: \beta_1 = 0.$

2. $H_a: \beta_1 \neq 0.$

3. Test statistic:

$$t = \frac{\hat{\beta}_1 - (0)}{s/\sqrt{SS_x}}.$$

4. Rejection region: Reject H_0 if $|t| > t_{\alpha/2}$ based on $n - 2$ degrees freedom.

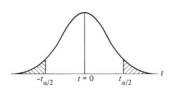

Example 10.4

For our data, test the hypothesis that there is no linear relationship between x and y at the $\alpha = .05$ level.

Solution

We will first need the quantity SS_x, which can be found from the data.

$$SS_x = \Sigma x_i^2 - \frac{(\Sigma x_i)^2}{n}$$

$$= 168 - \frac{(\underline{\hspace{2cm}})^2}{5}$$ 26

$$= 168 - \underline{\hspace{2cm}} = \underline{\hspace{2cm}}.$$ 135.2; 32.8

1. $H_0: \beta_1 = \underline{\hspace{2cm}}.$ 0

2. $H_a: \beta_1 \neq \underline{\hspace{2cm}}.$ 0

3. Test statistic:

$$t = \frac{\hat{\beta}_1 - (0)}{s/\sqrt{SS_x}}.$$

4. Rejection region: With 3 degrees of freedom, we shall reject H_0 if
$|t| > t_{.025} = \underline{\hspace{2cm}}.$ 3.182

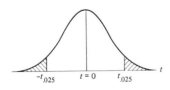

5. Calculate the statistic:

$$t = \frac{1.22 - 0}{.63/\sqrt{32.8}} = \frac{1.22}{\underline{\hspace{1.5cm}}} = \underline{\hspace{2cm}}.$$ .110; 11.09

3.182
reject; is

6. Since 11.09 is larger than the critical value of $t =$ _____ , we (reject, do not reject) H_0 and conclude that there (is, is not) a linear relationship between x and y.

Confidence Interval for β_1
If x increases one unit, what is the predicted change in y? Since $\hat{\beta}_1$ is an unbiased estimator for β_1 and has a normal distribution, the t-statistic, based on $n - 2$ degrees of freedom, can be used to derive the confidence interval estimator for the slope, β_1:

$$\hat{\beta}_1 \pm t_{\alpha/2} \frac{s}{\sqrt{SS_x}}.$$

Example 10.5
Find a 95% confidence interval for the average change in y for an increase of one unit in x.

Solution

.05; .025; 3
3.182

1. $1 - \alpha = .95$; $\alpha =$ _____ ; $\alpha/2 =$ _____ ; $n - 2 =$ _____ ;
$t_{.025} =$ _____ .

2. $\hat{\beta}_1 \pm t_{.025} \frac{s}{\sqrt{SS_x}}.$

3.182

$1.22 \pm ($ _____ $) \frac{.63}{\sqrt{32.8}}$

.35

$1.22 \pm ($ _____ $).$

.87; 1.57

3. A 95% confidence interval for β_1 is (_____ , _____).

Points Concerning Interpretation of Results

does not

If the test $H_0: \beta_1 = 0$ is performed and H_0 is *not rejected*, this (does, does not) mean that x and y are *not related*, since

II
linearly

1. a type _____ error may have been committed, or
2. x and y may be related, but not _____ . For example, the true relationship may be of the form $y = \beta_0 + \beta_1 x + \beta_2 x^2$.
If the test $H_0: \beta_1 = 0$ is performed and H_0 *is rejected*,

cannot

1. we (can, cannot) say that x and y are solely linearly related, since there may be other terms (x^2 or x^3) that have not been included in our model;
2. we should not conclude that a *causal* relationship exists between x and y, since the related changes we observe in x and y may actually be *caused* by an unmeasured third variable, say z.

Consider the problem where the true relationship between x and y is a "curve" rather than a straight line. Suppose we fitted a straight line to the data for values of x between a and b.

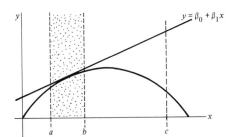

Using $\hat{y} = \hat{\beta}_0 + \hat{\beta}_1 x$ to predict values of y for $a \leqslant x \leqslant b$ would result in quite an accurate prediction. However, if the prediction line were used to predict y for the value $x = c$, the prediction would be highly _____. Although the line adequately describes the indicated trend in the region $a \leqslant x \leqslant b$, there is no justification for assuming that the line would fit equally well for values of x outside the region $a \leqslant x \leqslant b$. The process of predicting outside the region of experimentation is called _____. As our example shows, an experimenter should *not* extrapolate unless he is willing to assume the consequences of *gross errors*.

inaccurate

extrapolation

Self-Correcting Exercises 10B

1. Refer to Exercise 1, Self-Correcting Exercises 10A. Calculate *SSE*, s^2, and s for these data.
   a. Test the hypothesis that there is no linear relationship between actual and pre-enrollment figures at the $\alpha = .05$ level of significance.
   b. Estimate the average increase in actual enrollment for an increase of 100 in pre-enrolled students with a 95% confidence interval.
2. Refer to Exercise 2, Self-Correcting Exercises 10A. Calculate *SSE*, s^2, and s for these data and test for a significant linear relationship between yield and number of bolls at the .05 level of significance.
3. Refer to Exercise 3, Self-Correcting Exercises 10A. Test for a significant linear relationship between yield and the number of insects present at the $\alpha = .05$ level of significance.

10.6 Estimating the Expected Value of y for a Given Value of x

Assume that x and y are related according to the model

$$y = \beta_0 + \beta_1 x + \epsilon.$$

We have found an estimator for this line which is

$$\hat{y} = \hat{\beta}_0 + \hat{\beta}_1 x.$$

Suppose we are interested in estimating $E(y)$ for a given value of x, say x_0.

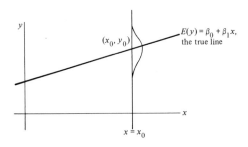

In repeated sampling, the predicted values of y will generate a distribution of estimates, $\hat{y}$, for the value of $x = x_0$. The mean of these estimates is the true value $E(y) = \beta_0 + \beta_1 x_0$. Therefore, we will use $\hat{y}$ to estimate the expected or average value of y for $x = x_0$, where the latter is written as

$$E(y|x_0) = \beta_0 + \beta_1 x_0.$$

$\beta_0 + \beta_1 x_0$

The estimator, $\hat{y} = \hat{\beta}_0 + \hat{\beta}_1 x_0$, has these properties:
1. $E(\hat{y}|x_0) = E(y|x_0)$. That is, for a fixed value of x, $x = x_0$, $\hat{y}$ is an unbiased estimator for the average value of y, $E(y|x_0) = \underline{\hspace{2cm}}$.
2. The variance of $(\hat{y}|x_0)$ is $\sigma_{\hat{y}}^2$ where

$$\sigma_{\hat{y}}^2 = \sigma^2 \left[\frac{1}{n} + \frac{(x_0 - \bar{x})^2}{SS_x} \right].$$

3. When the random component ϵ is normally and independently distributed, then $\hat{y}|x$ is normally distributed.

By using these results we can construct a z- or t-statistic to test an hypothesis concerning the expected value of y when $x = x_0$. Since σ^2 is rarely known, its sample estimate, s^2, is used, resulting in a $\underline{\hspace{2cm}}$-statistic with $\underline{\hspace{2cm}}$ degrees of freedom.

$t; n - 2$

Test of an Hypothesis Concerning $E(y/x_0)$

1. $H_0: E(y/x_0) = E_0.$

2. H_a: Appropriate one- or two-tailed test.
3. Test statistic:

$$t = \frac{\hat{y} - E_0}{\hat{\sigma}_{\hat{y}}} = \frac{\hat{y} - E_0}{s\sqrt{\dfrac{1}{n} + \dfrac{(x_0 - \bar{x})^2}{SS_x}}}.$$

4. Rejection region: Appropriate one- or two-tailed rejection region based on H_a.

Example 10.6
For our data, test the hypothesis that $\beta_0 = -1$ against the alternative that $\beta_0 < -1$ at the $\alpha = .05$ level.

Remark: By setting $x_0 = 0$, $E(y/x_0 = 0) = \beta_0 + \beta_1(0) = \beta_0$. Therefore, the test described above can be used to test an hypothesis about the intercept β_0.

Solution

1. $\quad H_0: E(y|x = 0) = \beta_0 = -1.$

2. $\quad H_a: \beta_0 < -1.$

3. Test statistic:

$$t = \frac{\hat{\beta}_0 - (-1)}{s\sqrt{\frac{1}{n} + \frac{(0 - \bar{x})^2}{SS_x}}}$$

$$= \frac{-1.34 - (-1)}{.63\sqrt{\frac{1}{5} + \frac{(0 - 5.2)^2}{32.8}}} = \frac{-.34}{\underline{\hspace{2cm}}} = \underline{\hspace{2cm}}.$$

.64; -.53

4. Rejection region: With _____ degrees of freedom, we shall reject H_0 if $t < $ _____.

3
-2.353

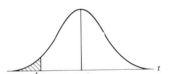

$t_{.05}$ $t = 0$

5. Decision: (Reject, Do not reject) H_0. The data (do, do not) present sufficient evidence to indicate that $\beta_0 < -1$.

Do not reject; do not

A $100(1 - \alpha)\%$ *confidence interval for* $E(y/x_0)$ *is given as*

$$(\hat{y}|x_0) \pm t_{\alpha/2} s\sqrt{\frac{1}{n} + \frac{(x_0 - \bar{x})^2}{SS_x}}$$

where $(\hat{y}|x_0)$ is the value of the estimate for $x = x_0$, found using

$$\hat{y} = \hat{\beta}_0 + \hat{\beta}_1 x_0.$$

Example 10.7
Find a 95% confidence interval for $E(y|x = 6)$.

Solution

1. $\quad (\hat{y}|x = 6) = -1.34 + 1.22(6) = $ _____.

5.98

2. $\quad t_{.025} = 3.182.$

.2952

3. $\hat{\sigma}_{\hat{y}} = (.63) \sqrt{\dfrac{1}{5} + \dfrac{(6 - 5.2)^2}{32.8}} = $ _____

4. A 95% confidence interval is constructed as follows:

$$(\hat{y}|x = 6) \pm 3.182\,\hat{\sigma}_{\hat{y}}$$

5.98; .2952

(_____) ± 3.182 (_____)

5.98; .94

(_____) ± (_____).

10.7 Predicting a Particular Value of y for a Given Value of x

In section 10.6 we attempted to estimate a point on the line $E(y) = \beta_0 + \beta_1 x$ when $x = x_p$. According to our probabilistic model, we actually observe points, y_i, which are deviations about the line

$$E(y) = \beta_0 + \beta_1 x$$

where the value y_i differs from $E(y_i)$ by the random amount ϵ_i.
We now wish to predict the value of a particular point, say y_i, when $x = x_p$. We have as a predictor for this value of y_i the quantity $\hat{y} = \hat{\beta}_0 + \hat{\beta}_1 x_p$.

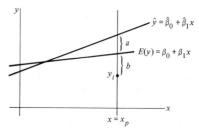

By considering the preceding graph, notice that the variance in predicting this value, y_p, has two components:
a. The variance in predicting $E(y|x_p)$, which is

$$\sigma^2 \left[\dfrac{1}{n} + \dfrac{(x_p - \bar{x})^2}{SS_x} \right]$$

b. The variance of the particular value of y from $E(y|x_p)$, which is σ^2. Hence the variance of the error in the prediction can be shown to be

$1 + \dfrac{1}{n} + \dfrac{(x_p - \bar{x})^2}{\Sigma\,SS_x}$

$$\sigma^2 + \sigma^2 \left[\dfrac{1}{n} + \dfrac{(x_p - \bar{x})^2}{\Sigma\,SS_x} \right] = \sigma^2 \,[\underline{\hspace{2cm}}].$$

When s^2 is used to estimate σ^2, a confidence interval based on the t-statistic can be constructed, resulting in the $(1 - \alpha)\,100\%$ prediction interval

$$(\hat{y}|x_p) \pm t_{\alpha/2}\, s\, \sqrt{1 + \frac{1}{n} + \frac{(x_p - \bar{x})^2}{SS_x}}\;.$$

Example 10.8

With 95% confidence, predict the particular value y_p when $x_p = 6$.

Solution

$\hat{y} = -1.34 + 1.22(6) = $ _____ .   5.98

With 3 degrees of freedom, $t_{.025} = $ _____ .   3.182
The prediction interval would be

$$5.98 \pm (\underline{\hspace{1cm}})\,(.63)\;\sqrt{1 + \frac{1}{5} + \frac{(6 - 5.2)^2}{32.8}}$$ 3.182

$$5.98 \pm (\underline{\hspace{1cm}}).$$ 2.21

Notice that the prediction interval for y_p is (wider, narrower) than the corresponding confidence interval for $E(y|x_p)$.   wider

Self-Correcting Exercises 10C

1. Refer to Exercise 1, Self-Correcting Exercises 10A. Test the hypothesis that the expected enrollment is zero if there are no students pre-enrolled at the $\alpha = .05$ level. Does the line of means pass through the origin?
2. For Exercise 2, Self-Correcting Exercises 10A, predict the expected yield in cotton when the mid-season boll count is 450 with a 90% confidence interval. Could you use the prediction line to predict the cotton yield if the mid-season count was 250?
3. Refer to Exercise 3, Self-Correcting Exercises 10A. Using the least-squares prediction line for these data, predict the expected cotton yield if the insect count is 12 with a 90% confidence interval. Compare this interval with that found in Exercise 2 and comment on these two predictors of cotton yield.
4. Use the least-squares line from Exercise 1, Self-Correcting Exercises 10A to predict the enrollment if the pre-enrollment figure is 4,000 students with 95% confidence.

10.8 A Coefficient of Correlation

The Pearson product moment coefficient of correlation is a measure of the strength of the _____ relationship between two variables and is   linear
independent of the respective scales of measurement. The sample coefficient of correlation used to estimate ρ, the population coefficient, is given as

$$r = \frac{\Sigma(x_i - \bar{x})(y_i - \bar{y})}{\sqrt{\Sigma(x_i - \bar{x})^2\,\Sigma(y_i - \bar{y})^2}}$$

where $-1 \leqslant r \leqslant 1$.

The computational formula for r is

$$r = \frac{SS_{xy}}{\sqrt{SS_x\,SS_y}}.$$

Examine the forms for r given above and note the following:

1. The denominator of r is the square root of the product of two positive quantities and will always be _____.

   positive

2. The numerator of r is identical to the numerator used to calculate _____, whose denominator is also always positive.

   $\hat{\beta}_1$

3. Hence _____ and r will always have the same algebraic sign. When

   $\hat{\beta}_1$

   $\hat{\beta}_1 > 0$, then r _____.

   > 0

   $\hat{\beta}_1 = 0$, then r _____.

   $= 0$

   $\hat{\beta}_1 < 0$, then r _____.

   < 0

When $r > 0$, there is a _____ linear correlation; when $r < 0$, there is a _____ linear correlation; when $r = 0$, there is _____ linear correlation. See the following examples:

positive

negative; no

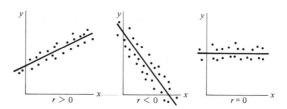

$r > 0$ $r < 0$ $r = 0$

Notice that both r and $\hat{\beta}_1$ measure the linear relationship between x and y. While r is independent of the scale of measurement, $\hat{\beta}_1$ retains the measurement units for both x and y, since $\hat{\beta}_1$ is the number of units increase in _____ for a one unit increase in _____. When should the investigator use r, and when should a least-squares estimate of β_1 be used? Although the situation is not always clear-cut, r is used when either x or y can be considered as the random variable of interest (i.e., when x and y are both random) while regression estimates and confidence intervals are appropriate when one variable, say x, is not random, and the other (y) is random.

$y; x$

Example 10.9

Two personnel evaluation techniques are available. The first requires a two-hour test-interview session while the second can be completed in less than an hour. A high correlation between test scores would indicate that the second test, which is shorter to use, could replace the two-hour test and hence save time and money. The following data give the scores on test I (x) and test II

(y) for $n = 15$ job applicants. Find the coefficient of correlation for the following pairs of scores:

Applicant	Test I (x)	Test II (y)
1	75	38
2	89	56
3	60	35
4	71	45
5	92	59
6	105	70
7	55	31
8	87	52
9	73	48
10	77	41
11	84	51
12	91	58
13	75	45
14	82	49
15	76	47

Solution:

1. As with a regression problem, we need all the summations as given in Section 10.4. Using a similar tabulation (or a calculator) find the following summations:

$$\Sigma x_i = \underline{\hspace{2cm}} \qquad \Sigma y_i = \underline{\hspace{2cm}} \qquad\qquad 1192; 725$$

$$\Sigma x_i^2 = 96990 \qquad\qquad \Sigma y_i^2 = 36461$$

$$\Sigma x_i y_i = \underline{\hspace{2cm}}. \qquad\qquad 59324$$

2. To use the computational formula for r we need:

$$SS_{xy} = \Sigma x_i y_i - \frac{(\Sigma x_i)(\Sigma y_i)}{n}$$

$$= 59324 - \frac{(\underline{\hspace{1.5cm}})(\underline{\hspace{1.5cm}})}{15} \qquad\qquad 1192; 725$$

$$= \underline{\hspace{2cm}} \qquad\qquad 1710.6667,$$

$$SS_x = \Sigma x_i^2 - \frac{(\Sigma x_i)^2}{n}$$

$$= 96990 - \frac{(\underline{\hspace{1.5cm}})^2}{15} \qquad\qquad 1192$$

$$= 2265.7333,$$

$$SS_y = \Sigma y_i^2 - \frac{(\Sigma y_i)^2}{n}$$

$$= 36461 - \frac{(\underline{\hspace{1.5cm}})^2}{15} \qquad\qquad 725$$

$$= 1419.3333.$$

then

$$r = \frac{SS_{xy}}{\sqrt{SS_x \, SS_y}}$$

1710.6667

$$= \frac{}{\sqrt{(2265.7333)\,(1419.3333)}}$$

1710.6667

$$= \frac{}{1793.1154}$$

.9540

$$= \underline{\hspace{2cm}}$$

3. Since the correlation coefficient has a maximum value of 1 and a minimum value of –1, it would appear that the correlation between these two test scores is quite strong, with the relationship being _____ linear as indicated by the sign of the correlation coefficient. In other words, a high score on test I would predict a _____ score on test II, or a low score on test I would predict a _____ score on test II.

positive

high
low

What more can be gleaned from knowing the value of r? How can one assess the strength of the linear relationship of two variables? If $r = 0$, it is fairly obvious that there appears to be no linear relationship between x and y. To evaluate non-zero values of r, let us consider two possible predictors of y.

If the x variate were not measured, we would be forced to use the model

$$y = \beta_0 + \epsilon$$

with $\hat{\beta}_0 = \bar{y}$, so that we would use

$$\hat{y} = \bar{y}$$

as the predictor of the response y. The sum of squares for error for this predictor would be

$$SSE_1 = \Sigma(y_i - \bar{y})^2 = SS_y .$$

Knowing the value of x as well as y, we would use the model

$$y = \beta_0 + \beta_1 x + \epsilon,$$

in which case the resulting sums of squares for error would be

$$SSE_2 = SS_y - \frac{[SS_{xy}]^2}{SS_x} .$$

If a linear relationship between x and y does exist, then SSE_1 would be larger than SSE_2.

A criterion for evaluating this relationship is to look at the ratio SSE_2/SSE_1, which can be simplified in the following manner:

1. Dividing SSE_2 by SSE_1 we obtain

$$\frac{SSE_2}{SSE_1} = \frac{SS_y}{SS_y} - \frac{[SS_{xy}]^2}{SS_x \, SS_y}$$

$$= 1 - r^2.$$

2. Rearranging this equation we find

$$r^2 = 1 - \frac{SSE_2}{SSE_1}$$

$$= \frac{SSE_1 - SSE_2}{SSE_1}.$$

Since the difference $SSE_1 - SSE_2$ represents a reduction in the sum of squares accomplished by using a linear relationship,

$r^2 =$ ratio of the reduction in the sum of squares achieved by using the linear model to the total sum of squares about the sample mean which would be used as a predictor of y if x were ignored.

A more understandable way of saying the same thing is to note that r^2 represents the amount of variability in y that is accounted for by knowing x. Thus we see that to evaluate a correlation coefficient r, we should examine r^2 to interpret the strength of the linear relationship between x and y. The quantity r^2 is often called the *coefficient of determination*.

   For our example, the value of r was found to be $r = .9540$; therefore, $r^2 =$ _____ . Hence, we have reduced the variability of our predictor by _____ by knowing the value of x.

.9101
91%

Self-Correcting Exercises 10D

1. Refer to Exercise 2, Self-Correcting Exercises 10A.
    a. Find the correlation between the number of bolls and the yield of cotton.
    b. Find the coefficient of determination, r^2, and explain its significance in using the number of cotton bolls to predict yield of cotton.
2. Refer to Exercise 3, Self-Correcting Exercises 10A.
    a. Find the value of r^2 and r for these data and explain the value of using the number of damaging insects present to predict cotton yield.
    b. Compare the values of r^2 using these two predictors of cotton yield. Which predictor would you prefer?
3. The data in Exercises 2 and 3, Self-Correcting Exercises 10A, are related in that for each field quadrate, the yield, the number of bolls, and the number of damaging insects were simultaneously recorded. Using this fact, calculate the correlation between the number of cotton bolls and the number of insects present for the 7 field quadrates. Does this value of r explain in any way the similarity of results using the predictor in Exercise 2 and that in Exercise 3?

EXERCISES

1. For the following equations (1) give the y-intercept, (2) give the slope, and (3) graph the line corresponding to the equation:

a. $y = 3x - 2$.

b. $2y = 4x$.

c. $-y = 0.5 + x$.

d. $3x + 2y = 5$.

e. $y = 2$.

2. The following data were obtained in an experiment relating the dependent variable, y (texture of strawberries), with x (coded storage temperature):

x	-2	-2	0	2	2
y	4.0	3.5	2.0	0.5	0.0

a. Find the least-squares line for the data.
b. Plot the data points and graph the least-squares line as a check on your calculations.
c. Calculate SSE, s^2, and s.
d. Do the data indicate that texture and storage temperature are linearly related? (Use $\alpha = .05$)
e. Predict the expected strawberry texture for a coded storage temperature of $x = -1$ with a 90% confidence interval.
f. Of what value is the *linear* model in increasing the accuracy of prediction as compared to the predictor, $\bar{y}$?
g. Estimate the particular value of y when $x = 1$ with a 98% confidence interval?
h. At what value of x will the width of the confidence interval for a particular value of y be a minimum, assuming n remains fixed?

3. In addition to increasingly large bounds on error, why should an experimenter refrain from predicting y for values of x outside the experimental region?

4. If the experimenter stays within the experimental region, when will the error in predicting a particular value of y be maximum?

5. What happens if the coefficient of linear correlation, r, assumes the value one?

6. An agricultural experimenter, investigating the effect of the amount of nitrogen (x) applied in 100 pounds per acre on the yield of oats (y) measured in bushels per acre, collected the following data:

x	1	2	3	4
y	22	38	57	68
	19	41	54	65

a. Fit a least-squares line to the data.
b. Calculate SSE and s^2.
c. Is there sufficient evidence to indicate that the yield of oats is linearly related to the amount of nitrogen applied? (Use $\alpha = .05$)

d. Predict the expected yield of oats with 95% confidence if 250 pounds of nitrogen per acre are applied.

e. Predict the average increase in yield for an increase of 100 pounds of nitrogen with 90% confidence.

f. Calculate r^2 and explain its significance in terms of predicting y, the yield of oats.

7. In an industrial process, the yield, y, is thought to be linearly related to temperature, x. The following coded data is available:

Temperature	0	0.5	1.5	2.0	2.5
Yield	7.2	8.1	9.8	11.3	12.9
	6.9	8.4	10.1	11.7	13.2

a. Find the least-squares line for this data.

b. Plot the points and graph the line. Is your calculated line reasonable?

c. Calculate SSE and s^2.

d. Do the data indicate a linear relationship between yield and temperature at the $\alpha = .01$ level of significance?

e. Calculate r, the coefficient of linear correlation, and interpret your results.

f. Calculate r^2 and interpret its significance in predicting the yield, y.

g. Test the hypothesis that $E(y|x = 1.75) = 10.8$ at the $\alpha = .05$ level of significance.

h. Predict the particular value of y for a coded temperature $x = 1$ with 90% confidence.

8. A horticulturist devised a scale to measure the viability of roses that were packaged and stored for varying periods of time before transplanting. y represents the viability measurement and x represents the length of time in days that the plant is packaged and stored before transplanting.

x	5	10	15	20	25
y	15.3	13.6	9.8	5.5	1.8
	16.8	13.8	8.7	4.7	1.0

a. Fit a least-squares line to the data.

b. Calculate SSE and s^2 for the data.

c. Is there sufficient evidence to indicate that a linear relationship exists between freshness and storage time? (Use $\alpha = .05$)

d. Estimate the mean rate of change in freshness for a one-day increase in storage time with a 98% confidence interval.

e. Predict the expected freshness measurement for a storage time of 14 days with 95% confidence.

f. Of what value is the linear model in preference to $\bar{y}$ in predicting freshness?

Chapter 11

ANALYSIS OF ENUMERATIVE DATA

11.1 The Multinomial Experiment

Examine the following experimental situations for any general similarities:
1. Two hundred people are classifed according to their blood type and the number of people in each blood type group is recorded.
2. A sample of one hundred items is randomly selected from a production line. Each item is classified as belonging to one of the three groups: acceptables, seconds, or rejects. The number in each group is recorded.
3. A random sample of fifty books is taken from the local library. Each book is assigned to one of the three categories: science, art, or fiction. The number of books in each category is recorded.

Each of these situations is similar to the others in that classes or categories are defined and the number of items falling into each category is recorded. Hence, these experiments result in enumerative or _____ data, **count** and have the following general characteristics which define the _____ **multinomial** experiment:
1. The experiment consists of n identical trials.
2. The outcome of each trial falls into one of k classes or cells.
3. The probability that the outcome of a single trial falls into cell i is p_i, $i = 1, 2, \ldots, k$, where p_i is _____ from trial to trial and **constant**

$$\sum_{i=1}^{k} p_i = \underline{\qquad}.$$

1

4. The trials are _____. **independent**
5. We are interested in $n_1, n_2, n_3, \ldots, n_k$, where n_i is the number of trials in which the outcome falls into cell i, and

$$\sum_{i=1}^{k} n_i = \underline{\qquad}.$$

n

2

p_2

n_2

np_2

The binomial experiment is a special case of the multinomial experiment. This can be seen by letting $k =$ _____ , and noting the following correspondences:

	Binomial	Multinomial (k = 2)
a.	n	n
b.	p	p_1
c.	q	
d.	y	_____ n_1
e.	$n - y$	
f.	$E(y) = np$	$E(n_1) = np_1$
g.	$E(n - y) = nq$	$E(n_2) =$ _____

For the multinomial experiment, we wish to make inferences about the associated population parameters, $p_1, p_2, \ldots, p_k$. A statistic that allows us to make inferences of this sort was developed by the British statistician, Karl Pearson, around 1900.

11.2 Pearson's Chi-Square Test.

For a multinomial experiment consisting of n trials with known (or hypothesized) cell probabilities, $p_i, i = 1, 2, \ldots, k$, we can find the expected number of items falling into the i-th cell by using

$$E(n_i) = np_i \qquad i = 1, 2, \ldots, k.$$

The cell probabilities are rarely known in practical situations. Consequently, we wish to estimate or test hypotheses concerning their values. If the hypothesized cell probabilities given are the correct values, then the *observed* number of items falling in each of the cells, n_i, should differ but slightly from the expected number, $E(n_i) = np_i$. Pearson's statistic (given below) utilizes the squares of the deviations of the observed from the expected number in each cell.

$$X^2 = \sum_{i=1}^{k} \frac{[n_i - E(n_i)]^2}{E(n_i)}$$

or

$$X^2 = \sum_{i=1}^{k} \frac{(n_i - np_i)^2}{np_i}.$$

Note that the deviations are divided by the expected number so that the deviations are weighted according to whether the expected number is large or small. A deviation of 5 from an expected number of twenty contributes $(5)^2/20 =$ _____ to X^2, while a deviation of 5 from an expected number of 10 contributes $(5)^2/10 =$ _____ or *twice* as much to X^2.

1.25

2.50

When n the number of trials is large, this statistic has an approximate x^2 distribution, provided the expected numbers in each cell are not too small. We will require as a rule of thumb that $E[n_i] \geqslant$ _____. This require-   5
ment can be satisfied by combining those cells with small expected numbers
until every cell has an expected number of at least _____. For small   5
deviations from the expected cell counts, the value of the statistic would be
(large, small), supporting the hypothesized cell probabilities. However, for   small
large deviations from the expected counts, the value of the statistic would be
(large, small), and the hypothesized values of the cell probabilities would be   large
_____. Hence, a one-tailed test is used, rejecting H_0 when X^2 rejected
is _____.   large

To find the critical value of x^2 used for testing, the degrees of freedom
must be known. Since the degrees of freedom change as Pearson's statistic
is applied to different situations, the degrees of freedom will be specified for
each application that follows. In general, the degrees of freedom are equal
to the number of cells less one degree of freedom for each independent linear
restriction placed upon the cell counts. One linear restriction that will
always be present is that

$$n_1 + n_2 + n_3 + \ldots + n_k = \underline{\qquad}.$$
     n

Other restrictions may be imposed by the necessity to estimate certain
unknown cell parameters or by the method of sampling employed in the
collection of the data.

11.3 A Test of an Hypothesis Concerning Specified Cell Probabilities

Let us consider the following problems concerning cell probabilities in a
multinomial experiment:

Example 11.1
Previous enrollment records at a large university indicate that of the total
number of persons that apply for admission, 60% are admitted unconditionally, 5% are admitted on a trial basis, and the remainder are refused
admission. Of 500 applications to date for the coming year, 329 applicants
have been admitted unconditionally, 43 have been admitted on a trial basis,
and the remainder have been refused admission. Do these data indicate a
departure from previous admission rates?

Solution
This experiment consists of classifying 500 applicants into one of three cells:
cell 1 (unconditional admission), cell 2 (conditional admission), and cell 3
(admission refused), where, under previous observation, $p_1 = .60$, $p_2 = .05$,
and $p_3 = .35$. The expected cell numbers are found to be

1. $E[n_1] = np_1 = 500(.60) = \underline{\qquad}.$ 300

2. $E[n_2] = np_2 = 500(\underline{\qquad}) = \underline{\qquad}.$ .05; 25

3. $E[n_3] = np_3 = \underline{\qquad}(\underline{\qquad}) = \underline{\qquad}.$ 500; .35; 175

•

Tabulating the results, we have

| | Admissions | | | |
	Unconditional	Conditional	Refused	Total
Observed	329	43	128	500
Expected	300	25	175	500

Using Pearson's statistic we can test the hypothesis that the cell probabilities remain as before, against the alternative hypothesis that at least one cell probability is different from those specified. We will compare the value of X^2 with a critical value of $\chi^2_{.05}$. The degrees of freedom are equal to the number of cells ($k = 3$) less one degree of freedom for the linear restriction, $n_1 + n_2 + n_3 = n = 500$. Therefore, the degrees of freedom are $k - 1 = 3 - 1 = 2$ and $\chi^2_{.05} = $ _____.

5.991

Formalizing this discussion we have the following statistical test of the enrollment data:

1. $H_0: p_1 = .60, p_2 = .05, p_3 = .35.$
2. H_a: At least one value of p_i is different from that specified by H_0.
3. Test statistic:

$$X^2 = \sum_{i=1}^{3} \frac{(n_i - np_i)^2}{np_i}.$$

5.991

4. Rejection region: Reject H_0 if $X^2 \geqslant \chi^2_{.05} = $ _____.

$$X^2 = \frac{(329 - 300)^2}{300} + \frac{(43 - 25)^2}{25} + \frac{(128 - 175)^2}{175}$$

$$= 2.803 + 12.96 + 12.62$$

$$= 28.383.$$

reject

5. Decision: $X^2 = 28.383 > 5.991$; hence, we (reject; do not reject) H_0. It appears that the admissions to date are not following the previously stated rates.

Some light can be shed on the situation by looking at the sample estimates of the cell probabilities. $\hat{p}_1 = 329/500 = .658, \hat{p}_2 = 43/500 = .086,$ and $\hat{p}_3$

.256

= _____. Notice that the percentage of unconditional admissions has risen slightly, the number of conditional admissions has increased, and the percentage refused admission has decreased at the expense of the first two categories. A final judgment would have to be made when admissions are closed and final figures in.

Example 11.2

A botanist performs a secondary cross of petunias involving independent factors controlling leaf shape and flower color where the factor "*A*" represents red color, "*a*" represents white color, "*B*" represents round leaves, and "*b*" represents long leaves. According to the Mendelian Model, the plants should exhibit the characteristics AB, Ab, aB, and ab in the ratio $9:3:3:1$. Of 160 experimental plants, the following numbers were observed: $AB, 95$; $Ab, 30$; $aB, 28$; $ab, 7$. Is there sufficient evidence to refute the Mendelian Model at the $\alpha = .01$ level?

Solution
Translating the ratios into proportions, we have

$$P(AB) = p_1 = 9/16.$$

$$P(Ab) = p_2 = 3/16.$$

$$P(aB) = p_3 = 3/16.$$

$$P(ab) = p_4 = 1/16.$$

The data are tabulated as follows:

Cell	AB	Ab	aB	ab
Expected	90	30	30	10
Observed	95	30	28	7

Perform a statistical test of the Mendelian Model using Pearson's statistic.
1. $H_0: p_1 = 9/16, p_2 = 3/16, p_3 = 3/16, p_4 = 1/16.$
2. $H_a: p_i \neq p_{i0}$ for at least one value of $i = 1, 2, 3, 4.$
3. Test statistic:

$$X^2 = \sum_{i=1}^{4} \frac{[n_i - E(n_i)]^2}{E(n_i)}.$$

4. Rejection region: With 3 degrees of freedom, we shall reject H_0 if

$$X^2 > \chi^2_{.01} = \underline{\hspace{2cm}}.$$ 11.3449

$$X^2 = \frac{(95 - 90)^2}{90} + \frac{(30 - 30)^2}{30} + \frac{(28 - 30)^2}{30} + \frac{(7 - 10)^2}{10}$$

$$= .2778 + .0000 + .1333 + .9000$$

$$= \underline{\hspace{2cm}}.$$ 1.3111

5. Decision: Since $X^2 = 1.3111 < 11.3449$, (reject, do not reject) H_0.   do not reject
   There (is, is not) sufficient evidence to refute the Mendelian Model.   is not

Self-Correcting Exercises 11A

1. A company specializing in kitchen products has produced a mixer in five different colors. A random sample of $n = 250$ sales has produced the following data:

Color	White	Copper	Avocado	Rose	Gold
Number Sold	62	48	56	39	45

Test the hypothesis that there is no preference for color at the $\alpha = .05$ level of significance. (Hint: if there is no color preference, then $p_1 = p_2 = p_3 = p_4 = p_5 = 1/5$.)

2. The number of Caucasians possessing the four blood types, A, B, AB, and O, are said to be in the proportions .41, .12, .03, and .44, respectively. Would the observed frequencies of 90, 16, 10, and 84, respectively, furnish sufficient evidence to refute the given proportions at the $\alpha = .05$ level of significance?

11.4 Contingency Tables

We now examine the problem of determining whether independence exists between two methods for classifying observed data. If we were to classify people first according to their hair color, and second according to their complexion, would these methods of classification be independent of each other? We might classify students first, according to the college in which they are enrolled, and second, according to their grade point average. Would these two methods of classification be independent? In each problem we are asking if one method of classification is *contingent* on another. We investigate this problem by displaying our data according to the two methods of classification in an array called a _____ table.

contingency

Example 11.3

A criminologist studying criminal offenders who have a record of one or more arrests, is interested in knowing if the educational achievement level of the offender influences the frequency of arrests. He has classified his data using four educational achievement level classifications:

A: completed 6th grade or less.
B: completed 7th, 8th or 9th grade.
C: completed 10th, 11th, or 12th grade.
D: education beyond 12th grade.

		Educational Achievement				
		A	B	C	D	Totals
	1	55 (45.39)	40 (43.03)	43 (43.03)	30 (36.55)	168
Number of Arrests	2	15 (21.61)	25 (20.49)	18 (20.49)	22 (17.40)	80
	3 or more	7 (10.00)	8 (9.48)	12 (9.48)	10 (8.05)	37
	Totals	77	73	73	62	285

The contingency table shows the number of offenders in each cell together with the expected cell frequency (in parentheses). The expected frequencies are obtained as follows:

1. Define p_A as the unconditional probability that a criminal offender will have completed grade 6 or less. Define $p_B, p_C,$ and p_D in a similar manner.

2. Define $p_1, p_2,$ and p_3 to be the unconditional probability that the offender has 1, 2, or 3 or more arrests, respectively.

If two events, A and B, are independent, then $P(AB) = $ _____. $P(A) \cdot P(B)$

Hence, if the two classifications are independent, a cell probability will equal the product of the two respective unconditional row and column probabilities. For example, the probability that an offender who has completed grade 6 is arrested 3 or more times is

$$p_{A3} = p_A \cdot p_3,$$

whereas the probability that a person with a 10th grade education is arrested twice is

$$p_{C2} = \underline{\hspace{2cm}}.$$ $p_C \cdot p_2$

Since the row and column probabilities are unknown, they must be estimated from the sample data. The estimators for these probabilities are defined in terms of r_i, the row totals, c_j, the column totals, and n.

1. $\hat{p}_A = c_1/n = 77/285.$
   $\hat{p}_B = c_2/n = 73/285.$
   $\hat{p}_C = c_3/n = 73/285.$
   $\hat{p}_D = c_4/n = 62/285.$
2. $\hat{p}_1 = r_1/n = 168/285.$
   $\hat{p}_2 = r_2/n = 80/285.$
   $\hat{p}_3 = r_3/n = 37/285.$

If the observed cell frequency for the cell in row i and column j is denoted by n_{ij}, then an estimate for the expected cell number in the ijth cell under the hypothesis of independence can be calculated by using the estimated cell probabilities.

$$E(n_{ij}) = n(p_{ij})$$

$$= n(p_i)\,(p_j).$$

$$\hat{E}(n_{ij}) = n(r_i/n)\,(c_j/n)$$

$$= r_i c_j/n.$$

The expected cell numbers enclosed in parentheses for the contingency table are found in this way. For example,

a. $\hat{E}(n_{11}) = \dfrac{(168)\,(77)}{285} = 45.39.$

b. $\hat{E}(n_{12}) = \dfrac{(168)\,(73)}{285} = 43.03.$

8.05

c. $\hat{E}(n_{34}) = \dfrac{(37)\,(62)}{285} = \underline{\hspace{2cm}}.$

80; 62; 17.40

d. $\hat{E}(n_{24}) = \dfrac{(\underline{\hspace{1.5cm}})\,(\underline{\hspace{1.5cm}})}{285} = \underline{\hspace{2cm}}.$

Now Pearson's statistic can be calculated accordingly as

$$X^2 = \sum_{i=1}^{3} \sum_{j=1}^{4} \frac{[n_{ij} - \hat{E}(n_{ij})]^2}{\hat{E}(n_{ij})}$$

$$= \frac{(55 - 45.39)^2}{45.39} + \frac{(40 - 43.03)^2}{43.03} + \ldots + \frac{(12 - 9.48)^2}{9.48}$$

$$+ \frac{(10 - 8.05)^2}{8.05}$$

$$= 10.23.$$

Recall that the number of degrees of freedom associated with the X^2-statistic equals the number of cells less one degree of freedom for each independent linear restriction on the cell counts. The first restriction is

n; one

that $\Sigma n_i = \underline{\hspace{2cm}}$; hence $\underline{\hspace{3cm}}$ degree of freedom is lost here. Then $(r - 1)$ independent linear restrictions have been placed on the cell counts due to the estimation of $(r - 1)$ row probabilities. Note that we need only estimate $(r - 1)$ independent row probabilities since their sum

(1) one

must equal $\underline{\hspace{3cm}}$. In like manner, $(c - 1)$ independent linear restrictions have been placed on the cell counts due to the estimation of the column probabilities.

Since there are rc cells, the number of degrees of freedom for testing X^2 in an $r \times c$ contingency table is

$$rc - (1) - (r - 1) - (c - 1),$$

which can be factored algebraically as

$$(r - 1)\,(c - 1).$$

In short, the number of degrees of freedom for an $r \times c$ contingency table, where all expected cell frequencies must be estimated from sample data (that is, from estimated row and column probabilities), is the number of rows minus one, times the number of columns minus one.

For the problem concerning criminal offenders, the degrees of freedom are

$$(r - 1)(c - 1) = (\underline{\hspace{1.5cm}})(\underline{\hspace{1.5cm}}) = \underline{\hspace{1.5cm}}.$$

2; 3; 6

We can now formalize the test of the hypothesis of independence of the two methods of classification at the $\alpha = .05$ level.

1. H_0: The two classifications are independent.
2. H_a: The two classifications are not independent.
3. Test statistic:

$$X^2 = \sum_{i=1}^{3} \sum_{j=1}^{4} \frac{[n_{ij} - \hat{E}(n_{ij})]^2}{\hat{E}(n_{ij})}.$$

4. Rejection region: With 6 degrees of freedom, we shall reject H_0 if $X^2 > \chi^2_{.05} = \underline{\hspace{1.5cm}}$.

12.5916

   The calculation of X^2 results in the value $X^2 = 10.23$.
5. Decision: Since $X^2 < 12.5916$, (do, do not) reject H_0. The data (do, do not) present sufficient evidence to indicate that educational achievement and the number of arrests are dependent.

do not
do not

Example 11.4

A sociologist wishes to test the hypothesis that the number of children in a family is independent of the family income. A random sample of 385 families resulted in the following contingency table:

Number of Children	Income Brackets in Thousands of Dollars				Total
	0–$4	$4–8	$8–12	Above $12	
0	10 (14.26)	9 (15.05)	18 (16.48)	24 (15.21)	61
1	8 (17.77)	12 (18.75)	25 (20.53)	31 (18.95)	76
2	14 (21.74)	28 (22.95)	23 (25.12)	28 (23.19)	93
3	26 (17.77)	24 (18.75)	20 (20.53)	6 (18.95)	76
4 or more	32 (18.47)	22 (19.49)	18 (21.34)	7 (19.70)	79
Total	90	95	104	96	385

If the number in parentheses is the estimated expected cell number, do these data present sufficient evidence at the $\alpha = .01$ level to indicate an independence of family size and family income?

Solution

The estimated cell counts have been found using

$$\hat{E}(n_{ij}) = \frac{r_i c_j}{n}$$

12

26.2170

Reject
are not

and are given in the parentheses within each cell. The degrees of freedom are $(r - 1)(c - 1) =$ _____.

1. H_0: The two classifications are independent.
2. H_a: The classifications are not independent.
3. Test statistic:

$$X^2 = \sum_{i=1}^{5} \sum_{j=1}^{4} \frac{[n_{ij} - \hat{E}(n_{ij})]^2}{\hat{E}(n_{ij})}.$$

4. Rejection region: With 12 degrees of freedom, we shall reject H_0 if

$$X^2 > \chi^2_{.01} = \text{_____}.$$

Calculate X^2:

$$X^2 = \frac{(10 - 14.26)^2}{14.26} + \frac{(9 - 15.05)^2}{15.05} + \ldots + \frac{(18 - 21.34)^2}{21.34}$$

$$+ \frac{(7 - 19.70)^2}{19.70}$$

$$= 63.4783.$$

5. Decision: (Reject, Do not reject) H_0. Therefore, we can conclude that family size and family income (are, are not) independent classifications.

Self-Correcting Exercises 11B

1. On the basis of the following data, is there a significant relationship between levels of income and political party affiliation at the $\alpha = .05$ level of significance?

Party Affiliation	Income		
	Low	Average	High
Republican	33	85	27
Democrat	19	71	56
Other	22	25	13

2. Three hundred people were interviewed to determine their opinions regarding a uniform driving code for all states.

Sex	Opinion	
	For	Against
Male	114	60
Female	87	39

Is there sufficient evidence to indicate that the opinion expressed is dependent upon the sex of the person interviewed?

11.5 $r \times c$ Tables with Fixed Row or Column Totals

To avoid having rows or columns that are absolutely empty, it is sometimes desirable to fix the row or column totals of a contingency table in the design of the experiment. In Example 11.4, the plan could have been to randomly sample 100 families in each of the four income brackets, thereby insuring that each of the income brackets would be represented in the sample. On the other hand, a random sample of 80 families in each of the "family size" categories could have been taken so that all "family size" categories would appear in the overall sample.

When using fixed row or column totals the number of independent linear restrictions on the cell counts is the same as for an $r \times c$ contingency table. Therefore, the data are analyzed in the same way that an $r \times c$ contingency table is analyzed, using Pearson's X^2 based on $(r - 1)(c - 1)$ degrees of freedom. In the following example we examine a case where the column totals are fixed in advance.

Example 11.5
Fifty 5th grade students from each of four city schools were given a standardized 5th grade reading test. After grading, each student was rated as satisfactory or not satisfactory in reading ability with the following results:

	School			
	1	2	3	4
Number Unsatisfactory	7	10	13	6

Is there sufficient evidence to indicate that the percentage of 5th grade students with an unsatisfactory reading ability varies from school to school?

Solution
The preceding table displays only half of the pertinent information. Extend the table to include the satisfactory category, allowing space to write in the expected cell frequencies.

	School				
	1	2	3	4	Total
Satisfactory	7 (9)	10 (____)	13 (____)	6 (____)	36
Not Satisfactory	43 (____)	40 (41)	37 (____)	44 (____)	164
Total	50	50	50	50	200

9; 9; 9

41; 41; 41

By fixing the column total at 50, we have made certain that the unconditional probability of observing a student from each of the schools is constant and equal to _____ for each school.

¼

If the percentage of unsatisfactory tests does not vary from school to school, then the probability of observing an unsatisfactory reading grade is the same for each school and equal to a common value p. Therefore, the unconditional probability of observing an unsatisfactory grade is p and in like manner, the probability of observing a satisfactory grade is $1 - p$

q $=$ _____ . If the percentage of unsatisfactory grades is the same for the four schools, then the probability of observing an unsatisfactory grade for a student in the jth school will be

$$p_{1j} = (¼)p \qquad \text{for } j = 1, 2, 3, 4,$$

while the probability of observing a satisfactory grade in the jth school will be

$$p_{2j} = (¼)q \qquad \text{for } j = 1, 2, 3, 4.$$

However, if the probability of an unsatisfactory grade varies from school to school, then

$$p_{1j} \neq (¼)p \quad \text{and} \quad p_{2j} \neq (¼)q$$

for at least one value of j where $j = 1, 2, 3, 4$. But this is the same as asking if the row and column classifications are independent; hence, the test is equivalent to a test of the independence of two classifications based upon $(r - 1)(c - 1)$ degrees of freedom.

Proceeding with the required test, we have

$$H_0 : p_1 = p_2 = p_3 = p_4 = p.$$

H_a: At least one proportion differs from at least one other.

Test statistic:

$$X^2 = \sum_i \sum_j \frac{[n_{ij} - \hat{E}(n_{ij})]^2}{\hat{E}(n_{ij})}.$$

3

7.81473

Rejection region: For $(2 - 1)(4 - 1) =$ _____ degrees of freedom, we shall reject H_0 if $X^2 > X^2_{.05} =$ _____ . To calculate the value of the test statistic we must first find the estimated expected cell counts.

9

$$\hat{E}(n_{11}) = \hat{E}(n_{12}) = \hat{E}(n_{13}) = \frac{(36)(50)}{200} = \text{_____}$$

41

$$\hat{E}(n_{21}) = \hat{E}(n_{22}) = \hat{E}(n_{23}) = \frac{(164)(50)}{200} = \text{_____}$$

$$X^2 = \frac{(7-9)^2}{9} + \frac{(10-9)^2}{9} + \frac{(\underline{\quad\quad})^2}{9} + \frac{(6-9)^2}{9}$$

13 - 9

$$+ \frac{(43-41)^2}{41} + \frac{(40-41)^2}{41} + \frac{(\underline{\quad\quad})^2}{41} + \frac{(44-41)^2}{41}$$

37 - 41

$$= \frac{\overline{}}{9} + \frac{\overline{}}{41} = 3.3333 + .7317 = \underline{\quad\quad}.$$

30; 30; 4.0650

Decision: Since $X^2 = 4.0650 < X^2_{.05} = 7.8147$, we (can, cannot) reject the hypothesis that reading ability for 5th graders as measured by this test does not vary from school to school.

cannot

Self-Correcting Exercises 11C

1. A survey of voter sentiment was conducted in four mid-city political wards to compare the fraction of voters favoring a "city-manager" form of government. Random samples of 200 voters were polled in each of the four wards with results as follows:

	Ward			
	1	2	3	4
Favor	75	63	69	58
Against	125	137	131	142

Can you conclude that the fractions favoring the city manager form of government differ in the four wards?

2. A personnel manager of a large company investigating employee satisfaction with their assigned jobs collected the following data for 200 employees in each of four job categories:

Satisfaction	Categories				
	I	II	III	IV	Total
High	40	60	52	48	200
Medium	103	87	82	88	360
Low	57	53	66	64	240
Total	200	200	200	200	

Do these data indicate that the satisfaction scores are dependent upon the job categories at the $\alpha = .05$ level?

11.5 Other Applications

The specific uses of the Chi-square test which we have dealt with in this chapter can be divided into two categories:

1. The first category is called "goodness of fit tests," whereby observed frequencies are compared with hyothesized frequencies which depend upon the hypothesized cell probabilities for a multinomial probability distribution. A decision is made as to whether the data fit the hypothesized model.

2. The second category is called "tests of independence," whereby a decision is made as to whether two methods of classifying the observations are statistically independent. If it is decided that the classifications are independent, then the probability that an observation would be classified as belonging to a specific row classification would be constant across the columns, or vice versa. If it is decided that the classifications are not independent, then the implication is that the probability that an observation would be classified as belonging to a specific row classification varies from column to column.

To illustrate the general nature of the goodness of fit test, we could test whether a set of data comes from any specified distribution such as the normal distribution with mean μ and variance σ^2, or a binomial distribution based upon n trials with probability of success p, or perhaps a Poisson distribution (we have not studied this distribution in any detail) with mean λ. Binomial data produce their own natural grouping corresponding to the cells of a multinomial experiment if one counts the number of zeros, ones, twos, and so on occurring in the data. If the expected cell frequencies are less than the required number, cells can be combined before using Pearson's statistic. Data from a normal distribution, on the other hand, do not produce an inherent natural grouping, and must be grouped as in a frequency histogram. In conjunction with a table of normal curve areas and the hypothesized normal distribution, the boundary points for the histogram should be chosen so that each "cell" has approximately the same probability and an expected

five

frequency greater than _____. Grouping the sample data accordingly one can compare the "observed" group frequencies against the theoretical ones using Pearson's Chi-square. If population parameters need to be

one

estimated, the point estimates given in earlier chapters are used and _____ degree of freedom subtracted for each independent estimate.

Tests of independence of two methods of classification are easily extended to three or more classifications by first estimating the expected cell frequencies and applying Pearson's statistic with the proper degrees of freedom. For example, in testing the independence of three classifications with c_1, c_2, and c_3 categories in the respective classifications, the test statistic would be

$$X^2 = \sum_{i=1}^{c_1} \sum_{j=1}^{c_2} \sum_{k=1}^{c_3} \frac{[n_{ijk} - \hat{E}(n_{ijk})]^2}{\hat{E}(n_{ijk})},$$

which has an approximate χ^2 distribution with $(c_1 - 1)(c_2 - 1)(c_3 - 1)$ degrees of freedom.

Further applications involving the χ^2 test are usually specifically tailored solutions to special problems. An example would be the test of a linear trend in a binomial proportion observed over time as discussed in your text.

Modifications such as this usually require the use of calculus and are beyond the scope of this text.

Self-Correcting Exercises 11D

1. A company producing wire rope has recorded the number of "breaks" occurring for a given type of wire rope within a four-hour period. These records were kept for 50 four-hour periods. If y is the number of "breaks" recorded for each four-hour period and μ is the mean number of "breaks" for a four-hour period, does the following Poisson Model adequately describe this data when $\mu = 2$?

$$p(y) = \frac{\mu^y e^{-\mu}}{y!} \qquad y = 0, 1, 2, \ldots$$

y	0	1	2	3 or more
Number Observed	4	15	16	15

Hint: Find $p(0), p(1)$, and $p(2)$ by means of Table 2 of your text. Use the fact that

$$P[y \geqslant 3] = 1 - p(0) - p(1) - p(2).$$

After finding the expected cell numbers, you can test the model by applying Pearson's Chi-square test.

2. In standardizing a score, the mean is subtracted and the result divided by the standard deviation. If 100 scores are so standardized and then grouped, test at the $\alpha = .05$ level of significance whether these scores were drawn from the standard normal distribution.

Interval	Frequency
Less than −1.5	8
−1.5 to −0.5	20
−0.5 to 0.5	40
0.5 to 1.5	29
Greater than 1.5	3

EXERCISES

1. What are the characteristics of a multinomial experiment?
2. Do the following situations possess the properties of a multinomial experiment?
   a. A large number of red, white, and blue flower seeds are thoroughly mixed and a sample of $n = 30$ seeds is taken. The numbers of red, white, and blue flower seeds are recorded.

b. A game of chance consists of picking three balls at random from an urn containing one white, three red, and six black balls. The game "pays" according to the number of white, red, and black balls chosen.

c. Four production lines are checked for defectives during an eight-hour period and the number of defectives for each production line recorded.

3. The probability of receiving grades of A, B, C, D, and E are .07, .15, .63, .10, and .05, respectively, in a certain humanities course. In a class of 120 students,

a. what is the expected number of A's?

b. what is the expected number of B's?

c. what is the expected number of C's?

4. A department store manager claims that his store has twice as many customers on Fridays and Saturdays as on any other day of the week (the store is closed on Sundays). That is, the probability that a customer visits the store Friday is 2/8, the probability that a customer visits the store Saturday is 2/8, while the probability that a customer visits the store on each of the remaining weekdays is 1/8. During an average week, the following numbers of customers visited the store:

Monday – 95	Thursday – 75
Tuesday – 110	Friday – 181
Wednesday – 125	Saturday – 214

Can the manager's claim be refuted at the $\alpha = .05$ level of significance?

5. If the probability of a female birth is 1/2, according to the binomial model, in a family containing four children, the probability of 0, 1, 2, 3, or 4 female births is 1/16, 4/16, 6/16, 4/16, and 1/16, respectively. A sample of 80 families each containing four children resulted in the following data:

Female Births	0	1	2	3	4
Number of Families	7	18	33	16	6

Do the data contradict the binomial model with $p = 1/2$ at the $\alpha = .05$ level of significance?

6. A serum thought to be effective in preventing colds was administered to 500 individuals. Their records for one year were compared to those of 500 untreated individuals with the following results:

	No Colds	One Cold	More Than One Cold
Treated	252	146	102
Untreated	224	136	140

Test the hypothesis that the two classifications are independent at the $\alpha = .05$ level of significance.

7. A manufacturer wished to know whether the number of defectives produced varied for four different production lines. A random sample of 100 items was selected from each line and the number of defectives recorded:

Production Lines	1	2	3	4
Defectives	8	12	7	9

Do these data produce sufficient evidence to indicate that the percentage of defects varies from line to line?

8. In a random sample of 50 male and 50 female undergraduates, each member was asked if he was for, against, or indifferent to the practice of having unannounced in-class quizzes. Do the following data indicate that attitude toward this practice is dependent upon the sex of the student interviewed?

	Male	Female
For	20	10
Against	15	30
Indifferent	15	10

9. In an experiment performed in a laboratory, a ball is bounced with a container whose bottom (or floor) has holes just large enough for the ball to pass through. The ball is allowed to bounce until it passes through one of the holes. For each of 100 trials, the number of bounces until the ball falls through one of the holes is recorded. If y is the number of bounces until the ball does fall through a hole, does the model

$$p(y) = (.6)(.4)^y \qquad y = 0, 1, 2, 3, \ldots$$

adequate describe the following data?

y	0	1	2	3 or more
Number Observed	65	28	4	3

Hint: First find $p(0), p(1), p(2)$, and $P(y \geq 3)$ from which the expected numbers for the cells can be calculated using np_0, np_1, np_2, and so on. Then a goodness of fit test will adequately answer the question posed.

Chapter 12

CONSIDERATIONS IN DESIGNING EXPERIMENTS

12.1 Introduction

In previous chapters we have used probabilistic concepts to develop methodologies for making inferences about populations in terms of their parameters. The emphasis in these chapters was on the analysis of data after the experiment had been conducted. Now that we have seen how information can be extracted from data, it is natural to ask (1) what factors affect the quantity of information in an experiment and (2) how can knowledge of these factors be used to design better experiments?

   The objective of this chapter is to acquaint the student with some of these factors and to show, in some simple cases, how the factors can be controlled in order to increase the amount of information that can be extracted from the data. The experiment should be designed to acquire a specified amount of information at a minimum cost.

12.2 Factors Affecting the Information in a Sample

There are two factors which affect the information in a sample available for making inferences about a population parameter:
1. The sample size (or sizes) used in experimentation.
2. The amount of variation in the experimental data.
Increasing the sample size or decreasing the amount of variation in the experiment tends to increase the amount of available information.

Example 12.1
Consider the problem of estimating a population mean μ with a $(1 - \alpha)$ 100% confidence interval, given as

$$\bar{y} \pm z_{\alpha/2} \frac{\sigma}{\sqrt{n}}.$$

The (wider, narrower) this confidence interval is, the more precise is the   narrower

estimate of μ and the more information we have gleaned from the experiment. Since $z_{\alpha/2}$ is a constant for fixed α, there are two ways in which the width of the confidence interval can be decreased:
1. Increase the sample size, n.
2. Decrease the population variance, σ^2.

Example 12.2
Consider the problem of estimating the difference between two population means with a $(1 - \alpha)$ 100% confidence interval. Give two ways in which the amount of available information can be increased.

Solution
The $(1 - \alpha)$ 100% confidence interval for $\mu_1 - \mu_2$ is

$(\bar{y}_1 - \bar{y}_2)$

$\pm z_{\alpha/2} \sqrt{\dfrac{\sigma_1^2}{n_1} + \dfrac{\sigma_2^2}{n_2}}$.

increased
decreased

_____ .

The width of the interval will be decreased if
1. one or both of the sample sizes are (decreased, increased).
2. one or both of the population variances are (decreased, increased).
The objective in the design of an experiment is to choose a method of increasing information (1. or 2.) which provides the specified amount of information at minimum cost.

12.3   The Physical Process of Designing an Experiment

Let us first consider some terminology used in experimental design so that the steps necessary in designing an experiment can be clearly defined.
1. The objects upon which measurements are taken are called *experimental units*.

Example 12.3
Identify the experimental units in the following situations:

can of peas

the person

rabbit

plot

    a. Twenty-five cans of peas are randomly selected from a production line and the weight of each can is recorded. _____

    b. Forty-three people are interviewed and their political affiliation is recorded. _____

    c. Seventeen rabbits, injected with a vaccine, are exposed to a virus and the number that become infected is recorded. _____

    d. In an agricultural experiment, a field is divided into plots, each of which will receive different amounts of fertilizer. The yield will then be recorded. _____

Quantitative
qualitative

2. Independent experimental variables are called *factors*, which may be either quantitative or qualitative. _____ factors can be measured and quantified, whereas _____ factors cannot.

Example 12.4
Designate the following factors as qualitative or quantitative:

Factor	Type	
a. amount of fertilizer	_____	quantitative
b. pressure in p.s.i.	_____	quantitative
c. political affiliation	_____	qualitative
d. time	_____	quantitative
e. distance	_____	quantitative
f. type of rope	_____	qualitative
g. model of car	_____	qualitative

3. The intensity setting of a factor is called a *level*, while a *treatment* is a specific combination of factor levels.

Example 12.5

Suppose that three types of steel cable are to be tested for tensile strength at three different temperature settings and four different pressure settings. This experiment would involve _____ qualitative factor(s) and _____ quantitative factor(s). The qualitative factor, types of cables, is set at _____ levels, while the quantitative factors, _____ and _____, are at _____ and _____ levels, respectively. There are _____ possible treatments corresponding to the (3) (3) (4) = 36 combinations of factor levels. One treatment would be the combination, "type I cable; first temperature setting; first pressure setting."

one
two
three
temperature; pressure; 3; 4
36

   Finally, we state the four steps necessary in designing an experiment:
1. Select the factors to be included in the experiment and specify the population parameters of interest.
2. Decide how much information is desired about the parameters of interest.
3. Select the treatments to be employed in the experiment and decide on the number of experimental units to be assigned to each.
4. Decide on the manner in which treatments should be applied to the experimental units.

12.4   Random Sampling and the Completely Randomized Design

Random sampling is a method of sampling which allows every possible sample in a population an equal chance of being selected. This method helps to avoid the possibility of bias introduced by a nonrandom selection of experimental units and provides a probabilistic basis for inference making. The random selection of independent samples from p populations is called a _____ _____ design.

completely randomized

   A simple and reliable method of random selection is found using the table of random numbers (Table 13, Appendix II). This table (and other random number tables) is constructed so that the ten integers occur randomly and with equal frequency.

   To find a starting place in the table, we could open our book to one of the pages in the table of random numbers and arbitrarily set our pencil point on the page. By reading to the right (left, up, or down) in groups of three digits, we stop at the first group of three digits between 001 and 200. These three digits will indicate the line in the table in which we are to begin.

Now we continue to move right (left, up, or down) in groups of two digits until a number between 01 and 14 is encountered. This number will indicate the column in which to begin. Suppose our pencil point came down on the number 9 in the 5-digit group 86902 in line 68 and column 7. Moving to the right, we find the next group of 5 digits so that we have

$$86902 \qquad 60397.$$

026
03

Hence, the starting line number is _____ and the starting column is _____ .

Example 12.6
An inspection plan requires that a random sample of ten radios be selected from an incoming lot containing 100 radios. We can consider the radios as being numbered 1 through 100, each radio numbered according to the order in which it was uncrated. Using the starting point found above, find the numbers of the ten radios to be included in the sample.

Solution
Beginning with the first group of five digits, 04839, one could move up or down, choosing groups of two digits unless the number 100 is encountered. In fact, any direction is permissible and the digits may be selected in any orderly manner. Lines 26–30 and columns 3–5 of Table 13, Appendix II are reproduced below:

04839	96423	24878
68086	26432	46901
39064	66432	84673
25669	26422	44407
64117	94305	26766

66; 94; 24

Using the first two digits in each group, we would include radios 4, 68, 39, 25, 64, 96, 26, _____ , _____ , and _____ in the sample. Note that the number 26 appeared twice in the random number table, but radio 26 can be used only once in the sample.

Example 12.7
Fifteen experimental units are to be used in a completely randomized design involving three treatments. If each treatment is to be applied to five units, find a randomization pattern for assigning the treatments to the experimental units.

Solution
Number the experimental units from 1 to 15. Using line 71 and column 9 in Table 13 as the starting point, using the first two digits in each line, and reading from top to bottom, the following numbers are obtained:

13; 05; 07; 11; 04

_____ , _____ , _____ , _____ , _____ .

From column 8, reading from bottom to top, we have

12, 09, 06, 14, 15.

These two sets of five numbers will designate the units to be assigned to the first two treatments. The remaining five numbers between 1 and 15 will designate the units to be assigned to the third treatment.

Treatment	Experimental Units
1	13, 5, 7, 11, 4
2	12, 9, 6, 14, 15
3	_____, _____, _____, _____, _____

1, 2, 3, 8, 10

Self-Correcting Exercises 12A

1. A large apartment complex houses 950 tenants. Use the random number table to identify the tenants to be included in a random sample of $n = 25$ tenants.

2. Find a randomization pattern for a completely randomized design for 20 experimental units and 4 treatments if each treatment is to be assigned to 5 experimental units.

3. An experimenter housed experimental animals two per cage. His experiment required that he randomly choose one of the animals in each cage as a control. His chance device was to reach into the cage and select as the animal to be treated, the one he was most easily able to catch.
   a. Will his selection be random? Why or why not?
   b. Use the random number table to devise a method of selection to replace his chance device.

12.5 Volume-Increasing Experimental Designs

Recall that there are two factors which influence the amount of information available in an experiment, the sample size and the variation in the experimental units. Consider now experiments which are designed to use the first factor to increase information through selection of factor level combinations and the number of experimental units assigned to each treatment.

There are many different "volume-increasing" designs, two of which have been discussed in the text. Optimal allocation of experimental units has been considered in the case of (1) comparing two population means and (2) estimating the slope parameter when fitting a straight line.

1. Consider the problem of estimating the difference in two population means. If the two population variances are equal and the experimental units of equal cost, the best allocation of n_1 and n_2 is to take $n_1 = n_2 = n$. If, however, the population variances are unequal, we should allocate n_1 and n_2 proportionally to their respective population standard deviations.

Example 12.8
Sixty experimental units are available to compare the means of two populations, A and B. What is the optimal allocation of the units if

a. $\sigma_A^2 = \sigma_B^2$?

b. $\sigma_A^2 = 100; \sigma_B^2 = 25$?

c. $\sigma_A^2 = 64; \sigma_B^2 = 4$?

Solution

a. For the case of equal variances, $n_A = n_B = n$ implies that $n_A = n_B$

30

$= \underline{\hspace{2cm}}$.

b. Since $\sigma_A = 10$ and $\sigma_B = 5$, twice as many units should be allocated to

40

population A as to population B. Hence, $n_A = \underline{\hspace{2cm}}$ and

20

$n_B = \underline{\hspace{2cm}}$.

c. Since $\sigma_A = 8$ and $\sigma_B = 2$, population A should receive 4 times as many

units. That is, one-fifth of the total will be allocated to population B

48

and four-fifths to population A. Hence, $n_A = \underline{\hspace{2cm}}$ and

12

$n_B = \underline{\hspace{2cm}}$.

2. Suppose we are interested in estimating the slope of the line, β_1, in the
linear model

$$y = \beta_0 + \beta_1 x + \epsilon$$

by the method of least squares. When we can control the n values of x
for which y will be observed, we should choose those values of x so that
the width of the confidence interval estimate will be a minimum. We seek
to choose x to minimize

$$\sigma_{\hat{\beta}_1} = \sqrt{\frac{\sigma^2}{\Sigma(x_i - \bar{x})^2}}.$$

decreases

Note that as $\Sigma(x_i - \bar{x})^2$ increases, $\sigma_{\hat{\beta}_1}$ (increases, decreases). The experimenter will specify the largest and smallest values of x which are of
interest to him, and it will remain for us to allocate the experimental units
over this range.

Example 12.9

In fitting a least-squares line, $n = 5$ measurements are to be taken with the
objective being the estimation of the slope, β_1, with a maximum amount of
information. The range of interest for x lies between 1 and 5. Compute $\sigma_{\hat{\beta}_1}$
for the following allocations:
1. Equal spacing of the 5 values of x.
2. Two points at each end and one in the center.

Solution

15

1. The five values of x are $1, 2, 3, 4$, and 5. Hence, $\Sigma x_i = \underline{\hspace{2cm}}$ and

55

$\Sigma x_i^2 = \underline{\hspace{2cm}}$ so that

10

$$\Sigma(x_i - \bar{x})^2 = \Sigma x_i^2 - \frac{(\Sigma x_i)^2}{5} = 55 - 45 = \underline{\hspace{2cm}}.$$

Then

$$\sigma_{\hat{\beta}_1} = \frac{\sigma}{\sqrt{10}}.$$

2. The five values of x are $1, 1, 3, 5,$ and 5 so that $\Sigma x_i =$ _____ and   15
   $\Sigma x_i^2 =$ _____. Then   61

$$\Sigma(x_i - \bar{x})^2 = 61 - \frac{225}{5} =$$ _____   16

and

$$\sigma_{\hat{\beta}_1} = \frac{\sigma}{\sqrt{16}}.$$

Note that $\sigma_{\hat{\beta}_1}$ is smaller for part $(1, 2)$ so that the (first, second) alloca-   2; second
tion represents a gain in information.
   In general, the smallest value of $\sigma_{\hat{\beta}_1}$ occurs when the n data points are
equally divided with half at the lower boundary of x and the other half at the
upper boundary of x. A few points should be selected near the center of the
experimental region to detect curvature if it is present.

Self-Correcting Exercises 12B

1. In comparing two population means with $\sigma_1^2 = 36$ and $\sigma_2^2 = 16$, what is
   the optimal allocation of $n = 50$ units between the two populations?
2. A chemist wishes to study the effect of temperatures between $120°F$ and
   $300°F$ on the reaction rate for a given chemical. He suspects that a linear
   relationship exists between temperature and rate of reaction and plans to
   run the experiment at 10 temperature settings.
   a. For equal spacings of temperature $(120°, 140°, \ldots 300°)$ what is the
      standard deviation of $\hat{\beta}_1$ if $\sigma = 10$?
   b. If five settings are taken at $120°$ and five at $300°$, calculate $\sigma_{\hat{\beta}_1}$ if
      $\sigma = 10$.
   c. How much larger is $\sigma_{\hat{\beta}_1}$ in a. than in b.?
   d. Which is the optimal allocation if the estimation of β_1 is of primary
      interest?

12.6 Noise-Reducing Experimental Designs

"Noise-reducing" designs increase the information in an experiment by
decreasing the variation in experimental units caused by uncontrolled
variables. This variation can be reduced by making all comparisons of
treatments within homogeneous groups of experimental units, called
blocks.

Example 12.10

Below are four examples of homogeneous groups or blocks of experimental units. In each case, specify the groups which constitute a block.

a. Determinations of potassium content in five different plants is made by a single laboratory technician. (_____ _____ _____)

determinations by the same technician

b. The yield for three varieties of wheat is measured in a field having no water or fertility trends. (_____ _____ _____ _____ _____)

plots of land having the same charac-teristics

c. The effect of two drugs upon manual dexterity is measured for a single subject. (_____ _____ _____ _____)

measurements on the same subject

d. For a given machine, the precision of the machine parts it produces is assessed. (_____ _____ _____ _____)

parts produced by the same machine

A randomized block design containing *p* treatments consists of *b* blocks of _____ experimental units each. The treatments are randomly assigned to the units in each _____ with each treatment appearing exactly _____ in every block.

p

block

once

Let us look at some situations where a block design can be used to reduce uncontrolled variation.

1. The potencies of several drugs are to be compared by three analysts. If each analyst makes one determination for each drug, the variability of these determinations should be homogeneous for a given analyst. Hence we can consider the _____ as blocks.

analysts

2. An experiment is to be conducted to assess the relative merits of five different gasolines. Since vehicle-to-vehicle variation is inevitable in such experiments, four vehicles are chosen and each of the five gasolines are used in each vehicle. In this case, we would take _____ as blocks.

vehicles

3. An experiment to assess the effects of three raw material suppliers and four different mixtures on the crushing strength of concrete blocks is to be run. In order to eliminate the variability from supplier to supplier, each of the four mixtures is prepared using the material from each of the three suppliers. Thus, we have reduced the variability by measuring the crushing strength of the four mixtures in each of three relatively homogeneous blocks, which are the _____ .

suppliers

Sometimes an identical trend, due to time or some other factor, may exist within each block. In such a case, blocking in two directions may be accom-plished by using a _____ _____ design. In such a design the horizontal blocks are referred to as _____ and the vertical block referred to as _____ . Each treatment appears exactly _____ in every row and every column; hence a Latin Square design requires that for *p* treatments, each row and each column contain _____ experi-mental units.

Latin Square

rows

columns; once

p

Example 12.11

In an experiment to assess the resistance to abrasion of three types of paint, a machine was used in which the samples could be placed in any of three positions. In addition, successive runs of this machine were known to yield

variable results. This experiment could be run as a _____ (give dimen-
sion) Latin Square using "positions" as columns, "runs" as _____
and "types of paint" as treatments.

3 × 3
rows

Complete the following design layout for this experiment:

Runs	Positions		
	1	*2*	*3*
1	B		
2	A	C	
3		B	A

A; C
B
C

Since each type of paint was abraded in each of the three positions and in
each of the three runs, the effect of positions and runs would be cancelled in
comparing the means.

Self-Correcting Exercises 12C

1. Explain how blocks could be used in the following situations to reduce
   experimental variation:
   a. Several brands of a food product are to be tested for volume of sales.
      The manufacturer expects to see a large difference in sales for several
      categories: (1) families with children, (2) senior citizens, and (3) young
      couples without children.
   b. The manager of a plywood manufacturing plant undertook a study to
      measure the bonding strength of four brands of glue. The experiment
      consisted of preparing sheets of three-quarter inch plywood from each
      of the four brands of glue. The bonding strength of the glue was tested
      for each of the sheets of plywood. The manager suspects that there may
      be variation due to the type of wood used in the plywood (fir, cedar,
      and redwood).
2. Complete the assignment of treatments for the following 4 × 4 Latin
   Square.

	A		
B	C		
		D	
A	D		

3. Suppose a field that is to be used in an agricultural experiment has an
   east-west water trend. A randomized block design could be used to
   eliminate the effect of the water trend.

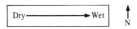

In what direction should the blocks be laid out and why?

EXERCISES

1. How is the width of a confidence interval related to the amount of information in a sample?

2. It has been stated in the text that two factors, "the magnitude of the background noise" and "the volume of the signal," affect the quantity of information in a sample. Give a short explanation as to what is meant by "noise" and "volume."

3. What is an experimental unit?

4. What is a factor?

5. Explain the difference between a quantitative and a qualitative factor.

6. Define the term treatment.

7. State the steps to be followed in designing an experiment.

8. Define a random sample and state two reasons for using a random sample.

9. Define a completely randomized design.

10. In a completely randomized design, 16 experimental units were to be equally divided between 2 treatments. Use the random number table to randomly assign the 16 units to the 2 treatments.

11. In a medical investigation, three stimulants were to be given to 7 subjects each. Use the random number table to randomly assign the 21 subjects to the three treatments.

12. Since the standard deviation of the estimator, $\hat{\beta}_1$, in the linear regression problem is minimized if the readings are equally divided, half run at the lowest setting and the other half run at the highest setting of x, is there any purpose in running any other points? If so, why and where should they be run?

Chapter 13

THE ANALYSIS OF VARIANCE

13.1 Introduction

Many investigations are directed toward establishing the effect of one or more variables upon a response of interest. The measurements that we record can be considered dependent variables, while their modifiers, be they treatments, classifications, or other factors, can be considered independent variables. A dependent variable, or response, is assumed to be a function of one or more independent variables which are varied and controlled by the experimenter during the investigation. These independent variables may be either qualitative or _____.

    In an investigation to determine how the amount of money invested in advertising affects the total sales, the _____ variable would be total sales and the _____ variable would be advertising expenditures. In this case the independent variable is (quantitative, qualitative).

    In the investigation of the effect of the birth order of an individual upon his or her social dominance rating, the dependent variable would be social dominance rating and the independent variable would be birth order. In this study, the independent variable is (qualitative, quantitative). In the last example, the investigator might also wish to consider the sex of the subject as an independent variable to be included in the study. Hence, there would be two independent variables, sex and birth order, with one dependent variable, social dominance rating.

    In this chapter we will use an analysis of variation as a statistical technique for evaluating the relationship between one or more independent variables, and a response, y.

13.2 The Analysis of Variance

To perform an analysis of variance is to partition the total variation in a set of measurements, given by

quantitative

dependent
independent
quantitative

qualitative

$$\sum_{i=1}^{n} (y_i - \bar{y})^2,$$

into portions associated with each independent variable in the experiment as well as a remainder attributable to random error. Let us investigate the partitioning of the total variation into two components with the following example.

Example 13.1
The impurities in parts per million were recorded for five batches of chemicals supplied by two different suppliers.

Supplier 1		Supplier 2	
	25		32
	33		43
	42		38
	27		47
	36		30
Sum	163	Sum	190
	$\bar{y}_1 = $ _____		$\bar{y}_2 = $ _____

32.6; 38

It will save confusion if we use two subscripts to identify each observation rather than just one. Let y_{ij} designate the jth observation recorded in the ith sample. When i is either 1 or 2, j can take the values 1, 2, 3, 4, or 5. We could then write:

Supplier 1	Supplier 2
$y_{11} = 25$	$y_{21} = 32$
$y_{12} = 33$	$y_{22} = 43$
$y_{13} = 42$	$y_{23} = 38$
$y_{14} = 27$	$y_{24} = 47$
$y_{15} = 36$	$y_{25} = 30$

1. *Total Variation:* Let us consider all measurements as one large sample of size 10. Then the total of the 10 measurements is

$$\sum_{i=1}^{2} \sum_{j=1}^{5} y_{ij} = 353$$

and the grand mean is

35.3

$$\bar{y} = \frac{353}{10} = \text{_____}.$$

The total variation then is given by

Total $SS = \sum_i \sum_j (y_{ij} - \bar{y})^2$

$$= \sum_i \sum_j y_{ij}^2 - \frac{\left(\sum_i \sum_j y_{ij}\right)^2}{10}$$

$$= 12929 - \frac{(353)^2}{10}$$

$$= 12929 - 12460.9$$

$$= \underline{\qquad}.$$

468.1

This total sum of squares will be partitioned into two sources of variation: treatments and error.

2. *Treatment Variation:* Recall that the variance of a sample mean is given to be σ^2/n, where n is the number in the mean and σ^2 is the variance in the population sampled. Suppose we had 2 samples of size n from the same population. Then, if $\bar{y}$ is the grand mean, the sample variance of the means,

$$s_{\bar{y}}^2 = \frac{\sum_{i=1}^{2} (\bar{y}_i - \bar{y})^2}{2 - 1}$$

estimates σ^2/n with $2 - 1 = 1$ degrees of freedom. If we multiply the sum of squares,

$$\sum_{i=1}^{2} (\bar{y}_i - \bar{y})^2$$

by n, we return this sum of squares to a "per measurement" basis. Then, the sum of squares due to variation of the treatment means will be

$$n \sum_{i=1}^{2} (\bar{y}_i - \bar{y})^2.$$

If the sample sizes are not equal, then the sum of squares for treatments is

$$SST = \sum_{i=1}^{2} n_i (\bar{y}_i - \bar{y})^2.$$

As the difference between the sample means increases, this sum of squares also _____. For the problem at hand, $p = 2$ and

increases

$$\bar{y}_1 = 32.6, \quad \bar{y}_2 = 38, \quad \bar{y} = 35.3$$

$$n_1 = n_2 = 5.$$

Therefore the treatment sum of squares is

$$
\begin{aligned}
SST &= n_1(\bar{y}_1 - \bar{y})^2 + n_2(\bar{y}_2 - \bar{y})^2 \\
&= 5(32.6 - 35.3)^2 + 5(38 - 35.3)^2 \\
&= 5(-2.7)^2 + 5(2.7)^2 \\
&= 5(7.29) + 5(7.29) \\
&= 2(36.45)
\end{aligned}
$$

72.9

$$= \underline{\qquad}.$$

3. *Error Variation:* If the two samples have come from the same population, we can use a pooled estimate of error given by

$$SSE = \sum_{j=1}^{5} (y_{1j} - \bar{y}_1)^2 + \sum_{j=1}^{5} (y_{2j} - \bar{y}_2)^2.$$

For sample 1,

$$
\begin{aligned}
\sum_{j=1}^{5} (y_{1j} - \bar{y}_1)^2 &= \sum_{j=1}^{5} y_{1j}^2 - \frac{(\Sigma y_{1j})^2}{5} \\
&= 5503 - \frac{(163)^2}{5} \\
&= 5503 - 5313.8
\end{aligned}
$$

189.2

$$= \underline{\qquad}.$$

For sample 2,

$$
\begin{aligned}
\sum_{j=1}^{5} (y_{2j} - \bar{y}_2)^2 &= \sum_{j=1}^{5} y_{2j}^2 - \frac{(\Sigma y_{2j})^2}{5} \\
&= 7426 - \frac{(190)^2}{5} \\
&= 7426 - 7220
\end{aligned}
$$

206

$$= \underline{\qquad}.$$

It follows that

$$SSE = 189.2 + 206$$

$$= \underline{\hspace{3cm}}.$$

395.2

4. Therefore we see directly that

$$SST = 72.9$$

$$SSE = 395.2$$

Total $SS = 468.1$

and that

Total $SS = SST + SSE$.

Since simpler calculational forms will be given presently, we defer further calculations until then.

The F Test and the Analysis of Variance

For the two-sample problem discussed in Example 13.1, the t-statistic is readily available for testing the hypothesis

$$H_0: \mu_1 = \mu_2$$

versus

$$H_a: \mu_1 \neq \mu_2$$

The two-sample unpaired t-test requires that both samples can be drawn randomly and independently from two normal populations with the same (equal) variances. With these assumptions we can construct an F-statistic, which is the ratio of two variances, to test these same hypotheses. The advantage to using the F-statistic is that the procedure can be easily extended for testing the equality of several population means.

If $H_0: \mu_1 = \mu_2$ is true, then the partitioning of the total sum of squares provides us with two estimators of the common variance, σ^2.

1. $$MSE = \frac{SSE}{n_1 + n_2 - 2}$$

where

$$SSE = \sum_{j=1}^{n_1} (y_{1j} - \bar{y}_1)^2 + \sum_{j=1}^{n_2} (y_{2j} - \bar{y}_2)^2$$

with $n_1 + n_2 - 2$ degrees of freedom.

2.
$$MST = \frac{SST}{2 - 1}$$

where

$$SST = n_1(\bar{y}_1 - \bar{y})^2 + n_2(\bar{y}_2 - \bar{y})^2$$

with $2 - 1$ degrees of freedom.
Therefore, when H_0 is true,

$$F = \frac{MST}{MSE}$$

$1; n_1 + n_2 - 2$

has an F distribution with $v_1 = $ _____ and $v_2 = $ _____ degrees of freedom.

MSE

If H_0 is false and $H_a : \mu_1 \neq \mu_2$ is true, this fact should be reflected in MST, and MST should in probability be larger than _____. This implies that if H_0 is false, the F-ratio,

$$F = \frac{MST}{MSE}$$

will be too large. Hence, the rejection region for this test will consist of all values of F satisfying

$$F > F_\alpha$$

where F_α is the right-tailed critical value of F having an area of α to its right based upon $v_1 = 1$ and $v_2 = n_1 + n_2 - 2$ degrees of freedom.

Example 13.2
For the data in Example 13.1, test the null hypothesis $H_0 : \mu_1 = \mu_2$ versus $H_a : \mu_1 \neq \mu_2$ at the $\alpha = .05$ level of significance.

Solution
Let us first gather the information that we have compiled so far.

$72.9; 72.9$

$$SST = \underline{\hspace{2cm}} \qquad MST = \frac{72.9}{1} = \underline{\hspace{2cm}}$$

$395.2; 49.4$

$$SSE = \underline{\hspace{2cm}} \qquad MSE = \frac{395.2}{8} = \underline{\hspace{2cm}}$$

1. $H_0 : \mu_1 = \mu_2$ \qquad versus \qquad $H_a : \mu_1 \neq \mu_2$.
2. The test statistic will be

$$F = \frac{MST}{MSE}$$

with v_1 = _____ and v_2 = _____ degrees of freedom.

3. Rejection region: For $\alpha = .05$, a right-tailed value of F with $v_1 = 1$ and $v_2 = 8$ degrees of freedom is $F_{.05}$ = _____. Therefore, we shall reject H_0 if $F >$ _____.

4. Using the sample values,

$$F = \frac{72.9}{49.4} = \underline{\qquad}.$$

1; 8

5.32
5.32

1.48

Since this value is (less, greater) than 5.32, we (reject, do not reject) the null hypothesis. There is not sufficient evidence to indicate that $\mu_1 \neq \mu_2$. Had we tested using the t-statistic,

less; do not reject

$$t = \frac{(\bar{y}_1 - \bar{y}_2) - 0}{\sqrt{s^2 \left(\dfrac{1}{n_1} + \dfrac{1}{n_2} \right)}}$$

with $n_1 + n_2 - 2 = 8$ degrees of freedom, the calculated value would have been

$$t = \frac{32.6 - 38.0}{\sqrt{49.4 \left(\dfrac{1}{5} + \dfrac{1}{5} \right)}}$$

$$= \frac{-5.4}{\sqrt{19.76}}$$

$$= \frac{-5.4}{4.445} = \underline{\qquad}.$$

-1.215

The rejection region would have consisted of values of t such that $|t| > t_{.025}$ = _____. Hence, we would not have rejected H_0: $\mu_1 = \mu_2$ even had we used the t-statistic. Noting that

2.306

$$t^2 = (-1.215)^2 = 1.48 = \underline{\qquad}$$

F

and

$$(t_{.025})^2 = (2.306)^2 = 5.32 = \underline{\qquad}$$

$F_{.05}$

the results should be identical. This verifies the fact that an F with $v_1 = 1$ and v_2 degrees of freedom is the same as t^2 with v_2 degrees of freedom; further, $F_{\alpha} = t^2_{\alpha/2}$ only if v_1 = _____.

1

13.3 The Analysis of Variance for Comparing More than Two Populations

In extending the problem of testing for a significant difference between two population means to one of testing for significant differences among several population means, let us consider an experiment run in a completely randomized design. A completely randomized design involves the selection of randomly drawn independent samples from each of p populations.

Consider an experiment designed to compare the growth of rats subjected to three specific diets, I, II, and III, for a given length of time. Fifteen rats have been randomly divided into three groups of five and each group randomly assigned to receive one of the diets. Since each rat is "treated" or subjected to one of the three diets, in statistical terminology the diets are called _____. This type of randomization procedure is called a _____ _____ design.

treatments

completely randomized

The completely randomized design involves one independent variable, treatments. In this design, the total sum of squares of deviations of the measurements about their overall mean can be partitioned into two parts.

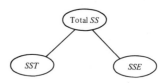

In generalizing the two-sample problem, we will now assume that *the p samples have been randomly and independently drawn from p normal populations with means* $\mu_1, \mu_2, \ldots, \mu_p$ *respectively, and with common variance,* σ^2.

Let T_i and $\overline{T}_i$ be the sum and the mean of the n_i observations in the sample from the ith population, with $n = n_1 + n_2 + \ldots + n_p$ being the total number of observations. Then

1.
$$\text{Total } SS = \sum_{i=1}^{p} \sum_{j=1}^{n_i} (y_{ij} - \overline{y})^2$$

n - 1

with _____ degrees of freedom.

2.
$$SST = \sum_{i=1}^{p} n_i (\overline{T}_i - \overline{y})^2$$

p - 1

with _____ degrees of freedom.

3.
$$SSE = \sum_{i=1}^{p} \sum_{j=1}^{n_i} (y_{ij} - \overline{T}_i)^2$$

with $\displaystyle\sum_{i=1}^{p} (n_i - 1) = n_1 + n_2 + \ldots + n_p - p = $ _____ degrees of

$n - p$

freedom. Not only does

$$\text{Total } SS = SST + SSE$$

but the same relationship holds for the degrees of freedom associated with each sum of squares.

$$d.f._{\text{Total}} = d.f._{\text{Treatments}} + d.f._{\text{Error}}$$

since

$$n - 1 = (p - 1) + (n - p).$$

The formulas actually used for computing these sums of squares are given below. Let

$$CM = \frac{\left(\displaystyle\sum_{i=1}^{p}\sum_{j=1}^{n_i} y_{ij}\right)^2}{n} = \frac{(\text{Grand Total})^2}{n}.$$

msp 349 summations

Then

1. $\displaystyle\text{Total } SS = \sum_{i=1}^{p}\sum_{j=1}^{n_i} y_{ij}^2 - CM.$

summations

2. $\displaystyle SST = \sum_{i=1}^{p} \frac{T_i^2}{n_i} - CM.$

3. $SSE = \text{Total } SS - SST.$

Notice that in SST, the square of each treatment total is divided by the number of _____ in that total. Although SSE can be computed directly as a pooled sum of squared deviations within each sample, it is computationally easier to use the additivity property, $SST + SSE$

observations

= _____.

Total SS

The mean squares for treatments and error are calculated by dividing each sum of squares by its degrees of freedom. Therefore

$$s^2 = MSE = \frac{SSE}{(\underline{\qquad})}$$

$n - p$

and

p - 1

$$MST = \frac{SST}{(\underline{\hspace{2cm}})}.$$

If all samples are from the same normal population, then MST and MSE are each estimators of the population variance, σ^2. In this case, the statistic

$$F = \frac{MST}{MSE}$$

F

has an _____ distribution with $v_1 = p - 1$ and $v_2 = n - p$ degrees of freedom.

Consider testing the hypothesis $H_0: \mu_1 = \mu_2 = \ldots = \mu_p$ against the alternative that at least one mean is different from at least one other. *If H_0 is true, then all samples have come from the same normal population* and the statistic

$$F = \frac{MST}{MSE}$$

has the F distribution specified above. However, if H_a is true (at least one of the equalities does not hold), then

$$MST = \frac{1}{p-1} [n_1(\bar{T}_1 - \bar{y})^2 + n_2(\bar{T}_2 - \bar{y})^2 + \ldots + n_p(\bar{T}_p - \bar{y})^2]$$

larger

larger; large

will in probability be _____ than MSE and F will tend to be _____ than expected. Hence, H_0 will be rejected for _____ values of F; that is, we shall reject H_0 if

$$F > F_\alpha$$

with $v_1 = p - 1$ and $v_2 = n - p$ degrees of freedom.

Example 13.3
Do the following data provide sufficient evidence to indicate a difference in the means of the three underlying treatment populations?

	Treatment		
	1	*2*	*3*
	3	7	5
	4	9	4
	2	8	5
		7	

9; 31; 14

T_i	_____	_____	_____	
				Total = 54
n_i	3	4	3	$n = 10$
$\bar{T}_i$	3	7.75	4.67	

Solution

1. We must first partition the total variation into *SST* and *SSE*.

$$CM = \frac{(54)^2}{10} = \frac{2916}{10} = \underline{\hspace{2cm}}.$$

291.6

a.   Total $SS = 3^2 + 4^2 + 2^2 + \ldots + 5^2 + 4^2 + 5^2 - 291.6$

$$= 338 - 291.6$$

$$= \underline{\hspace{2cm}}.$$

46.4

b. $SST = \dfrac{9^2}{3} + \dfrac{31^2}{4} + \dfrac{14^2}{3} - 291.6$

$$= \frac{81}{3} + \frac{961}{4} + \frac{196}{3} - 291.6$$

$$= 27 + 240.25 + 65.33 - 291.6$$

$$= \underline{\hspace{2cm}} - 291.6$$

332.58

$$= \underline{\hspace{2cm}}.$$

40.98

c. $SSE = 46.4 - 40.98$

$$= \underline{\hspace{2cm}}.$$

5.42

d. To compute the degrees of freedom we need the values $n = 10, p = 3$.
   Hence, *SST* has $p - 1 = \underline{\hspace{2cm}}$ degrees of freedom while   2
   SSE has $n - p = \underline{\hspace{2cm}}$ degrees of freedom.   7
   The resulting mean squares are

$$MST = \frac{40.98}{2} = \underline{\hspace{2cm}}$$

20.49

$$MSE = \frac{5.42}{7} = \underline{\hspace{2cm}}.$$

0.77

2. We are now in a position to test $H_0: \mu_1 = \mu_2 = \mu_3$ versus H_a: at least
   one equality does not hold.
   a. The test statistic is

$$F = \frac{MST}{MSE}$$

with $v_1 = \underline{\hspace{2cm}}$ and $v_2 = \underline{\hspace{2cm}}$ degrees of freedom.   2; 7

4.74

b. The rejection region: Using $\alpha = .05$, we shall reject H_0 if $F > F_{.05}$
= _____ .

c. For our example,

26.61

$$F = \frac{20.49}{0.77} = \underline{\hspace{2cm}},$$

greater; reject

which is (greater, less) than $F_{.05} = 4.74$. Hence we _____
H_0 and conclude that there is evidence to indicate a difference in
means for the three treatment populations at the $\alpha = .05$ level of
significance.

Example 13.4
The length of time required for kindergarten age children to assemble a device
was compared for four different lengths of pre-experiment instructional
times. Four students were randomly assigned to each group but two were
eliminated during the experiment due to sickness. The data (assembly times
in minutes) are shown below:

	Pre-Experiment Instructional Time (hours)			
	.5	1.0	1.5	2.0
	8	9	4	4
	14	7	6	7
	9	5	7	5
	12		8	
T_i	43	21	25	16
$\overline{T}_i$	10.75	7.00	6.25	5.33
n_i	4	3	4	3

Do the data present sufficient evidence to indicate a difference in mean time
to assemble the device for the four different lengths of instructional time?
(Use $\alpha = .01$.)

Solution

1. $\quad CM = \dfrac{(105)^2}{14} = 787.50.$

107.5

a. $\quad$ Total $SS = 895.0 - 787.5 = \underline{\hspace{2cm}}.$

b. $\quad SST = \dfrac{43^2}{4} + \dfrac{21^2}{3} + \dfrac{25^2}{4} + \dfrac{16^2}{3} - CM = 63.33.$

44.17

c. $\quad SSE = 107.5 - 63.33 = \underline{\hspace{2cm}}.$

d. $\quad$ With $n = 14$ and $p = 4$,

$$MST = \frac{SST}{p-1} = \frac{63.33}{3} = \underline{\hspace{2cm}}.$$ $\qquad$ 21.11

$$MSE = \frac{SSE}{n-p} = \frac{44.17}{10} = \underline{\hspace{2cm}}.$$ $\qquad$ 4.42

2. Test of the null hypothesis.
   a. $H_0: \mu_1 = \mu_2 = \mu_3 = \mu_4$.
   b. H_a: At least one equality does not hold.
   c. Test statistic:

$$F = \frac{MST}{MSE}$$

   with $v_1 = \underline{\hspace{2cm}}$ and $v_2 = \underline{\hspace{2cm}}$ degrees of freedom. $\qquad$ 3; 10
   d. Rejection region: Reject H_0 if $F > F_{.01} = \underline{\hspace{2cm}}$. $\qquad$ 6.55
   e. For these data,

$$F = \frac{MST}{MSE} = 4.78.$$

   Hence, we (reject, do not reject) H_0; we (can, cannot) conclude that $\qquad$ do not reject; cannot
   sufficient evidence exists to indicate a difference in the mean time to
   assemble for the four levels of pre-experiment instructional time.

13.4 The Analysis of Variance Summary Table

The results of an analysis of variance are usually displayed in an analysis of
variance ($ANOVA$ or AOV) summary table. The table displays the sources
of variation together with the degrees of freedom, sums of squares, and
mean squares for each source listed in the table. The results of the F test
appear as a final entry in the table. For a completely randomized design, the
$ANOVA$ table is as follows:

ANOVA				
Source	d.f.	SS	MS	F
Treatments	$p-1$	SST	MST	
Error	$n-p$	SSE	MSE	
Total	$n-1$	Total SS		

$\qquad$ MST/MSE

This display gives all the pertinent information leading to the F test and
further emphasizes the fact that the degrees of freedom and the sums of
squares are both additive.

Example 13.5
Display the results of the analysis of the data in Example 13.3.

Solution

We need but collect the results that we have for this example.

ANOVA

Source	d.f.	SS	MS	F
Treatments	_____	40.98	20.49	_____
Error	_____	5.42	_____	
Total	_____	_____		

Margin: 2; 26.61 · 7; 0.77 · 9; 46.40

Example 13.6

Display the results of the analysis of the data in Example 13.4.

Solution

Since the term "treatments" is a general way of describing the differences in the sampled populations, we can replace the term "treatments" in this problem by the more descriptive word "times."

ANOVA

Source	d.f.	SS	MS	F
Times	3	63.33	_____	_____
Error	10	44.17	_____	
Total	_____	107.50		

Margin: 21.11; 4.78 · 4.42 · 13

13.5 Estimation for the Completely Randomized Design

In using the analysis of variance F-test to test for significant differences among a group of population means, an experimenter can conclude that either (a) there is no difference among the means or (b) at least one mean is different from at least one other. In the second case, (b), an experimenter may wish to proceed with estimating the value of a treatment mean or with estimating the difference between two treatment means.

Since the analysis of variance requires that all samples be drawn from _____ *(normal)* populations with a common variance, confidence intervals can be constructed using the t-statistic with error degrees of freedom. Hence, for estimating the ith treatment mean with a $(1 - \alpha)$ 100% confidence interval, use

$$\bar{T}_i \pm t_{\alpha/2} \sqrt{\frac{MSE}{n_i}},$$

where $\bar{T}_i$ is the ith sample mean, n_i is the number of observations in the ith sample, and MSE is the pooled estimate of σ^2 from the analysis of variance with _____ *(n – p)* degrees of freedom.

To estimate the difference between two population means with a $(1 - \alpha)$ 100% confidence interval, use

$$(\bar{T}_i - \bar{T}_j) \pm t_{\alpha/2} \sqrt{MSE \left[\frac{1}{n_i} + \frac{1}{n_j} \right]}.$$

Example 13.7
Refer to Example 13.4. Estimate the mean time to assemble for treatment 1 with a 95% confidence interval.

Solution
The required information can be obtained from Example 13.4 and Example 13.6.

$\bar{T}_1 = 10.75$　　　$d.f. = 10 = (n - p) = 14 - 4$

$n_1 = 4$　　　$t_{.025} = \underline{\hspace{1cm}}.$　　　2.228

$MSE = 4.42$

Therefore, the estimate is given by

$$10.75 \pm 2.228 \sqrt{\frac{4.42}{4}}$$

$10.75 \pm 2.228\,(\underline{\hspace{1cm}})$　　　1.05

$10.75 \pm \underline{\hspace{1cm}}$　　　2.34

or

$$\underline{\hspace{1cm}} < \mu_1 < \underline{\hspace{1cm}}.$$　　　8.41; 13.09

Example 13.8
Refer to Example 13.4. Construct a 95% confidence interval for $\mu_1 - \mu_2$.

Solution
Collect pertinent information.

$\bar{T}_1 = 10.75$　　$\bar{T}_2 = 7.0$　　$MSE = 4.42$　　$t_{.025} = 2.228$

$n_1 = 4$　　$n_2 = 3$　　$d.f. = 10. = (n - p) = 14 - 4$

The confidence interval is found using

$$(\bar{T}_1 - \bar{T}_2) \pm t_{.025} \sqrt{MSE \left[\frac{1}{n_1} + \frac{1}{n_2} \right]},$$

which when evaluated becomes

$$(10.75 - 7.0) \pm 2.228 \sqrt{4.42 \left(\frac{1}{4} + \frac{1}{3}\right)}$$

3.75; 3.58

$$\underline{\hspace{2cm}} \pm \underline{\hspace{2cm}}.$$

0.17

7.33

Hence, with 95% confidence, we estimate that $\mu_1 - \mu_2$ lies between \underline{\hspace{1.5cm}} and \underline{\hspace{1.5cm}} minutes.

Self-Correcting Exercises 13A

1. In the evaluation of three rations fed to chickens grown for market, the dressed weights of five chickens fed from birth on one of the three rations were recorded.

	Rations		
	1	2	3
	7.1	4.9	6.7
	6.2	6.6	6.0
	7.0	6.8	7.3
	5.6	4.6	6.2
	6.4	5.3	7.1
Total	32.3	28.2	33.3
Averages	6.46	5.64	6.66

a. Do the data present sufficient evidence to indicate a difference in the mean growth for the three rations as measured by the dressed weights?

b. Estimate the difference in mean weight for rations 2 and 3 with a 95% confidence interval.

2. In the investigation of a citizens' committee complaint about the availability of fire protection within the county, the distance in miles to the nearest fire station was measured for each of 5 randomly selected residences in each of four areas.

	Areas				
	1	2	3	4	
	7	1	7	4	
	5	4	9	6	
	5	3	8	3	
	6	4	7	7	
	8	5	8	5	
T_i	31	17	39	25	Total = 112
n_i	5	5	5	5	n = 20
$\bar{T}_i$	6.2	3.4	7.8	5.0	

a. Do these data provide sufficient evidence to indicate a difference in mean distance for the four areas at the $\alpha = .01$ level of significance?

b. Estimate the mean distance to the nearest fire station for those residents in area 1 with a 95% confidence interval.

c. Construct a 95% confidence interval for $\mu_1 - \mu_3$.

13.6 The Randomized Block Design

The randomized block design is a natural extension of the _____ _____ experiment. The purpose is to increase the _____ in the design by making comparisons between treatments within relatively homogeneous blocks of experimental material. The randomized block design for p treatments and b blocks assumes blocks of relatively homogeneous material with each block containing _____ experimental units. Each treatment is applied to one experimental unit in each block. Consequently, the number of observations for a given treatment for the entire experiment will equal _____. Thus, for the randomized block design, $n_1 = n_2 = \ldots = n_p = b$. A randomized block design for $p = 3$ treatments and $b = 4$ blocks is shown below. Denote the treatments as $T_1, T_2,$ and T_3.

paired
difference; information

p

b

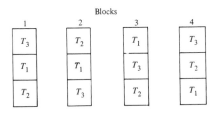

The total number of observations for a randomized block design with b blocks and p treatments is $n =$ _____.

The word "randomized" means that the _____ are randomly distributed over the experimental units within each block. The randomized block design involves two independent variables: _____ and _____. For an experiment run in a randomized block design, the total variation can now be partitioned into three sources of variation; blocks (B), treatments (T), and error (E).

bp
treatments

blocks
treatments

Randomized block designs prove to be very useful since many investigations involve human (or animal) subjects which exhibit a large subject-to-subject variability.
1. By using a subject as a "_____" and having each subject receive all the treatments in a random order, treatment comparisons made within subjects would exhibit less variation than treatment comparisons made between subjects.
2. Since every subject receives each treatment in some random order, a _____ number of subjects would be required in a randomized block design than in a completely randomized design.

block

smaller

13.7 The Analysis of Variance for a Randomized Block Design

In partitioning the sums of squares for an experiment run in a randomized block design with b blocks and p treatments, the calculational formulas for Total SS and SST remain the same except that now every treatment total will contain exactly b measurements and the total number of observations will be $n = pb$. Hence,

$$CM = \frac{\left(\displaystyle\sum_{i=1}^{p} \sum_{j=1}^{b} y_{ij} \right)^2}{pb} = \frac{(\text{Grand Total})^2}{pb}$$

$$\text{Total } SS = \sum_{i=1}^{p} \sum_{j=1}^{b} y_{ij}^2 - CM$$

and

$$SST = \sqrt{\frac{\displaystyle\sum_{i=1}^{p} T_i^2}{b}} - CM.$$

The calculation of SSB follows the same pattern as the calculation of SST, namely, square each block total; sum the squares of each total; divide by p, the number of observations per total; subtract the correction for the mean. If $B_1, B_2, \ldots, B_b$ represent the block totals, then

$$SSB = \frac{(B_1^2 + B_2^2 + \ldots + B_b^2)}{p} - CM.$$

Using the additivity of the sums of squares, we can find SSE by subtraction.

SSB; SST

$$SSE = \text{Total } SS - \underline{\hspace{3cm}} - \underline{\hspace{3cm}}.$$

The analysis of variance table for a randomized block design with b blocks and p treatments follows.

	ANOVA		
Source	*d.f.*	*SS*	*MS*
Blocks	_____	SSB	$SSB/(b-1)$
Treatments	_____	SST	$SST/(p-1)$
Error	_____	SSE	$SSE/(b-1)(p-1)$
Total	$bp - 1$	Total SS	

(b – 1)
(p – 1)
(b – 1) (p – 1)

To test the null hypothesis: "there is no difference in treatment means," we use

$$F = \frac{MST}{MSE},$$

which has an F distribution with $v_1 = p - 1$ and $v_2 = (p - 1)(b - 1)$ degrees of freedom when H_0 is true. If H_0 is false, the statistic will tend to be larger than expected; hence we would reject H_0 if

$$F > F_\alpha.$$

Although a test of H_0: "there is no difference in block means" is not always required, we can test this hypothesis using

$$F = \frac{MSB}{MSE}.$$

When H_0 is true, this statistic has an F distribution with $v_1 = b - 1$ and $v_2 = (b - 1)(p - 1)$ degrees of freedom. H_0 is rejected if $F > F_\alpha$, where F_α is an α-level critical value of F with $(b - 1)$ and $(b - 1)(p - 1)$ degrees of freedom.

A significant test of block means provides a method of assessing the efficiency of the experimenter's blocking procedure, since if

$$F = \frac{MSB}{MSE}$$

is significant, the experimenter has _____ the available informa-   **increased**
tion in the experiment by blocking, and would certainly use this same technique in subsequent experiments. In situations where subjects are often used as blocks, a nonsignificant test of block means should not be taken as a license to discontinue blocking in subsequent experiments involving different subjects, since the next group of subjects selected for participation could exhibit strong subject-to-subject variability and provide a highly significant test of blocks.

Example 13.9
The readability of four different styles of textbook types was compared using a speed-reading test. The amount of reading material was identical for all four type styles. The sample material for each of the four type styles was read in random order by each of five readers in order to eliminate the natural variation in reading speed between readers. The length of time to completion of reading was measured. Thus, each reader corresponds to a _____ and comparisons of the four styles were made within   **block**
readers. Do the data present evidence of a difference in mean reading times for the four type styles?

Type Style			Readers			Totals	Means
	1	2	3	4	5		
1	15	18	13	21	15	82	16.40
2	19	19	16	22	15	91	18.20
3	13	20	14	21	16	84	16.80
4	11	18	12	17	12	70	14.00
Totals	58	75	55	81	58	327	
Means	14.50	18.75	13.75	20.25	14.50		

Solution

Before analyzing these data, it should be pointed out that the order in which the type style was presented was randomized for each reader. The data layout presented *does not* represent the order of presentation for each reader.

1. *Partitioning the Sums of Squares*

$$CM = \frac{(\text{Total})^2}{bp} = \frac{(327)^2}{20} = 5346.45.$$

$$\text{Total } SS = \sum_{i=1}^{p} \sum_{j=1}^{b} y_{ij}^2 - CM$$

5346.45

$$= 5555 - \underline{\qquad}$$

208.55

$$= \underline{\qquad}.$$

For *SST* and *SSB* remember that the respective totals are each squared and divided by the number of measurements per total. Block totals contain $p = 4$ measurements and treatment totals contain $b = 5$ measurements.

$$SSB = \sum_{j=1}^{b} \frac{B_j^2}{p} - CM$$

5346.45

$$= \frac{58^2 + 75^2 + \ldots + 58^2}{4} - \underline{\qquad}$$

$$= 5484.75 - 5346.45$$

138.30

$$= \underline{\qquad}.$$

$$SST = \sum_{i=1}^{p} \frac{T_i^2}{b} - CM$$

$$= \frac{82^2 + 91^2 + 84^2 + 70^2}{5} - \underline{\qquad}$$

5346.45

$$= \underline{\qquad} - 5346.45$$

5392.20

$$= \underline{\qquad}.$$

45.75

$$SSE = Total\ SS - SSB - SST$$

$$= 208.55 - 138.30 - 45.75$$

$$= \underline{\qquad}.$$

24.50

2. *The Analysis of Variance Table*
Complete the following *ANOVA* table.

		ANOVA		
Source	d.f.	SS	MS	F
Blocks	4	138.30	_____	_____
Treatments	3	45.75	_____	_____
Error	12	24.50	_____	
Total	19	208.55		

34.58; 16.95
15.25; 7.48
2.04

In testing H_0: "no difference in treatment means," we use

$$F = \frac{MST}{MSE} = \frac{15.25}{2.04} = \underline{\qquad}.$$

7.48

With $v_1 = 3$ and $v_2 = 12$ degrees of freedom, $F_{.05} = \underline{\qquad}$. Hence we reject H_0 and conclude that there is a significant difference among the mean reading times for the four type styles.

3.49

Because we expected significant differences in mean reading times for the five readers, we used a randomized block design with the readers as blocks. Let us test whether these five readers have significantly different mean reading times. To test H_0: "no difference in block means," we use

$$F = \frac{MSB}{MSE} = \frac{34.58}{2.04} = \underline{\qquad}.$$

16.95

The 5% critical value of F with $v_1 = 4$ and $v_2 = 12$ degrees of freedom is _____; hence we _____ H_0 and conclude that the mean reading times for the five readers are significantly different. A significant test of blocks indicates that our experiment has been made (more, less) precise by using the randomized block design with readers as blocks.

3.26; reject

more

13.8 Estimation for the Randomized Block Design

Since a randomized block design involves two classifications, not only can we estimate differences in treatment means, but we can also estimate the differences in block means. In either situation we can construct a confidence interval estimate based upon Student's t distribution with _____ degrees of freedom. Hence $(1 - \alpha)$ 100% confidence intervals would be found using

$(b - 1)(p - 1)$

$$(\bar{T}_i - \bar{T}_j) \pm t_{\alpha/2} \sqrt{\frac{2\ MSE}{b}}$$

and

$$(\bar{B}_i - \bar{B}_j) \pm t_{\alpha/2} \sqrt{\frac{2\ MSE}{p}}.$$

Example 13.10
Refer to Example 13.9. Estimate the difference in mean reading time for type styles 1 and 2 with a 95% confidence interval.

Solution
Pertinent information:

2.04

$\bar{T}_1 = 16.40 \qquad MSE =$ _____

12

$\bar{T}_2 = 18.20 \qquad d.f. \ =$ _____

2.179

$b \ = 5 \qquad t_{.025} =$ _____.

Using

$$(\bar{T}_1 - \bar{T}_2) \pm t_{.025} \sqrt{\frac{2\ MSE}{b}}$$

we have

$$(16.40 - 18.20) \pm 2.179 \sqrt{\frac{2(2.04)}{5}}$$

or

−1.80; 1.97

_____ $\pm$ _____ minutes.

Example 13.11
For the same problem, find a 95% confidence interval for the difference in mean reading time for readers 2 and 3.

Solution

To use the estimator

$$(\bar{B}_2 - \bar{B}_3) \pm t_{.025} \sqrt{\frac{2\,MSE}{p}}$$

we need the following additional information: $\bar{B}_2 =$ _____,
$\bar{B}_3 =$ _____, $p = 4$. Then

$$(18.75 - 13.75) \pm 2.179 \sqrt{\frac{2(2.04)}{4}}$$

18.75

13.75

simplifies to

_____ $\pm$ _____ minutes.

5.00; 2.20

   Confidence interval estimates for block means or the difference between two block means are not always required nor are they always useful. Often, however, blocks constitute an important factor in many experiments. For example, if a marketing research survey investigating potential sales for four products is carried out in six areas, the areas would constitute blocks for this experiment and the differences in mean sales between areas would be of strong interest to the experimenter.

Self-Correcting Exercises 13B

1. An experiment was conducted to compare four feed additives on the growth of pigs. To eliminate genetic variability, pig litters were used as blocks. Five litters were employed, with four pigs selected randomly from each litter. Each group of four pigs, tending to be more homogeneous than those between litters, was considered a block. The data (growth in pounds) is shown below.

		Additive		
Litter	1	2	3	4
1	78	69	78	85
2	66	64	70	70
3	81	78	72	83
4	76	66	77	74
5	61	66	69	70

   a. Do the data present sufficient evidence to indicate a difference in mean growth for the four additives?
   b. Do the data present sufficient evidence to indicate a difference in mean growth between litters? Was blocking desirable?
   c. Find a 95% confidence interval for the difference in mean growth for additives 1 and 2.
   d. Find a 95% confidence interval for the mean growth for additive 3.

2. Example 9.6 was analyzed in Chapter 9 as a paired difference experiment. The data have been reproduced below.

Pair	Conventional	New
1	78	83
2	65	69
3	88	87
4	91	93
5	72	78
6	59	59

a. Analyze the data to detect a difference in means using the analysis of variance for a randomized block design. (Use $\alpha = .05$.)
b. What is the relationship between the calculated value of t (Example 9.6) and the calculated value of F?

13.9 The Analysis of Variance for a Latin Square Design

The Latin Square design is a further extension of the randomized block design. In the randomized block design we attempted to increase the information in the experiment by isolating from the total variation the variation due to differences in blocks, and by making comparisons between treatments within relatively homogeneous blocks of experimental material. The Latin Square design is used when a third source of variation, in addition to blocks and treatments, is present. For example, in an agricultural field experiment, the field that is to be used may have an east-west fertility trend, and at the same time a north-south water trend due to the contour of the land.

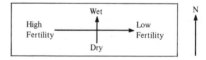

To eliminate the effects due to the water and fertility trends, we can lay out
blocks in _____ directions, east to west and north to south. The
blocks running east to west will be relatively homogeneous with respect to
the _____ trend, while blocks running north to south will be
relatively homogeneous with respect to the _____ trend. We refer
to the east-west blocks as _____, and the north-south blocks
as _____.

two

water
fertility
rows
columns

The Latin Square design requires that the number of rows and columns equal the number of treatments. In addition, each treatment must be applied to one experimental unit in each row and column. Hence, for a $p \times p$ Latin Square, each treatment will appear p times in the experiment, and there will be p different treatments. A Latin Square design for $p = 4$ treatments is shown below. The total number of observations is $p^2 = 16$.

Columns

	1	2	3	4
1	T_1	T_2	T_3	T_4
2	T_4	T_1	T_2	T_3
3	T_3	T_4	T_1	T_2
4	T_2	T_3	T_4	T_1

Rows

The Latin Square design involves three independent variables: rows, columns, and treatments. For an experiment run in such a design, the total variation can be partitioned into four sources of variation: rows (R), columns (C), treatments (T), and error (E), so that

$$\text{Total } SS = SSR + SSC + SST + SSE.$$

In partitioning the sum of squares, the formulas for Total SS and SST are calculated in the same manner as for the completely randomized or randomized block design. Each treatment total will contain p measurements, and the total number of observations will be p^2. In addition,

1.
$$SSR = \sum_{i=1}^{p} \frac{R_i^2}{p} - CM, \text{ and}$$

2.
$$SSC = \sum_{j=1}^{p} \frac{C_j^2}{p} - CM,$$

where R_i is the total for row i and C_j is the total for column j. The correction for the mean is calculated as before. Finally,

$$SSE = \text{Total } SS - SSR - SSC - SST.$$

An analysis of variance table for a Latin Square design is shown below.

ANOVA for Latin Square Design

Source	d.f.	SS	MS
Rows	$p - 1$	SSR	$SSR/(p-1)$
Columns	$p - 1$	SSC	$SSC/(p-1)$
Treatments	$p - 1$	SST	$SST/(p-1)$
Error	$(p-1)(p-2)$	SSE	$SSE/(p-1)(p-2)$
Total	$p^2 - 1$	Total SS	

F tests to test the three hypotheses,

1. H_0: No difference among treatments

2. H_0: No difference among rows

3. H_0: No difference among columns

are, respectively,

1. $$F = \frac{MST}{MSE}$$

2. $$F = \frac{MSR}{MSE}$$

3. $$F = \frac{MSC}{MSE}$$

If H_0 is true, the necessary test statistic has an F distribution with $v_1 = p - 1$ and $v_2 = (p - 1)(p - 2)$ degrees of freedom. H_0 will be rejected if

$$F > F_\alpha,$$

where F_α is the critical value of F with area α to its right and v_1 and v_2 degrees of freedom.

Example 13.12
An experiment was conducted to investigate the effects of four different display techniques used to improve sales. The experiment was run in a Latin Square design utilizing four different stores as columns and four different shopping days as rows. The following responses represent sales in hundreds of dollars for the display techniques, $A, B, C,$ and D.

Store	1	2	3	4	Totals	Means
Day 1	24 (B)	30 (D)	46 (C)	12 (A)	112	28
2	42 (D)	4 (A)	12 (B)	32 (C)	90	22.5
3	8 (A)	8 (C)	36 (D)	28 (B)	80	20
4	28 (C)	30 (B)	28 (A)	54 (D)	140	35
Totals	102	72	122	126	422	
Means	25.5	18.0	30.5	31.5	26.38	

	A	B	C	D
Treatment totals:	52	94	114	162
Treatment means:	13	23.5	28.5	40.5

Is there evidence to indicate a difference in mean sales for the four display techniques?

Solution
1. *Partitioning the sums of squares*

$$CM = \frac{(\text{Total})^2}{p^2} = \frac{442^2}{16} = 11130.25.$$

Total $SS = \sum_i \sum_j y_{ij}^2 - CM$

$= 14276 - 11130.25 = \underline{\hspace{2cm}}$. | 3145.75

$$SSR = \sum_{i=1}^{4} \frac{R_i^2}{p} - CM$$

$= 11661 - 11130.25 = \underline{\hspace{2cm}}$. | 530.75

$$SSC = \sum_{i=1}^{4} \frac{C_j^2}{p} - CM$$

$= 11587 - 11130.25 = \underline{\hspace{2cm}}$. | 456.75

$$SST = \sum_{i=1}^{4} \frac{T_i^2}{p} - CM$$

$= 12695 - 11130.25 = \underline{\hspace{2cm}}$. | 1564.75

2. *The Analysis of Variance Table*

		ANOVA			
Source	d.f.	SS	MS	F	
Rows (days)	_____	530.75	_____	_____	3; 176.92; 1.79
Columns (stores)	_____	456.75	_____	_____	3; 152.25; 1.54
Treatments (techniques)	_____	1564.75	_____	_____	3; 521.58; 5.27
Error	_____	593.50	_____		6; 98.92
Total	15	3145.75			

To test the null hypothesis "no difference in treatment means" we use

$$F = \frac{MST}{MSE} = \frac{521.58}{98.92} = \underline{\hspace{2cm}}.$$ | 5.27

The critical value of F for $\alpha = .05$, based on $v_1 = 3$ and $v_2 = 6$ degrees of freedom is _____. Since the computed value of F is (less, greater) than the critical value, we (can, cannot) reject the null hypothesis of no treatment differences; that is, there does appear to be a significant difference in mean sales for the 4 display techniques.

| 4.76; greater
| can

To test the hypothesis of "no difference in mean sales for days," we use

176.92; 1.79

$$F = \frac{MSR}{MSE} = \frac{(\underline{\hspace{1cm}})}{98.92} = \underline{\hspace{1cm}}.$$

cannot

does not

The critical value of F for an .05 significance level is, again, 4.76. We find that we (can, cannot) reject the null hypothesis of "no row differences"; that is, there (does, does not) appear to be a significant difference in sales from day to day.

One further test, the test of the null hypothesis, "there is no difference in mean sales for stores," can be performed using

152.25; 1.54

$$F = \frac{MSC}{MSE} = \frac{(\underline{\hspace{1cm}})}{98.92} = \underline{\hspace{1cm}}.$$

cannot

Using the same critical value of F, we (can, cannot) reject the null hypothesis, since 1.54 is less than the critical value, 4.76.

13.10 Estimation for the Latin Square Design

No change is encountered in the construction of a $(1 - \alpha)$ 100% confidence interval for the difference in two treatment means, row means, or column means. These confidence interval estimators are given respectively as

$$(\bar{T}_i - \bar{T}_j) \pm t_{\alpha/2} \, s \sqrt{\frac{2}{p}}.$$

$$(\bar{R}_i - \bar{R}_j) \pm t_{\alpha/2} \, s \sqrt{\frac{2}{p}}.$$

$$(\bar{C}_i - \bar{C}_j) \pm t_{\alpha/2} \, s \sqrt{\frac{2}{p}}.$$

Example 13.13
Construct a 95% confidence interval for the difference in mean sales for display techniques A and D.

Solution
From the data table and the *ANOVA* table, we see that

13.0; 40.5

$$\bar{T}_A = \underline{\hspace{1cm}}, \bar{T}_D = \underline{\hspace{1cm}}, s^2 = MSE = 98.92, \text{ and}$$

9.95

$$s = \underline{\hspace{1cm}}.$$

2.447

Based on 6 degrees of freedom, $t_{.025}$ is $\underline{\hspace{1cm}}$. Hence the 95% confidence interval is given as

$(13.0 - 40.5) \pm 2.447\,(9.95)\,(\underline{\hspace{2cm}})$ | $\sqrt{.5}$

or

$$\underline{\hspace{2cm}} \pm \underline{\hspace{2cm}}.$$ | $-27.5;\,17.21$

Therefore, we estimate the difference in mean sales for A and D to lie between 10.29 and 44.71, or between $1,029 and $4,471.

Example 13.14
Construct a 95% confidence interval for the difference in mean sales for stores 1 and 2.

Solution

$$\bar{C}_1 = 25.5 \qquad \bar{C}_2 = 18.00 \qquad s = 9.95 \qquad t_{.025} = 2.447.$$

The 95% confidence interval is

$$(25.5 - 18) \pm 2.447\,(9.95)\,\sqrt{.5}$$

$$\underline{\hspace{2cm}} \pm \underline{\hspace{2cm}}.$$ | $7.5;\,17.21$

Hence, we estimate the mean difference in sales for the two stores to be between _____ and _____, which is to say, the estimated mean difference in sales will be between $-$971 and $2,471. | $-9.71;\,24.71$

Self-Correcting Exercises 13C

1. An experiment was conducted to compare the yield of five different varieties of corn, A, B, C, D, and E. Because a fertility and a water trend were present, a Latin Square design was used. In the following table the response given is the number of bushels of corn harvested per experimental unit.

			Columns			
Rows	1	2	3	4	5	Totals
1	25 (A)	36 (B)	36 (C)	30 (D)	34 (E)	161
2	24 (B)	27 (C)	17 (D)	18 (E)	24 (A)	110
3	29 (C)	24 (D)	29 (E)	28 (A)	36 (B)	146
4	30 (D)	34 (E)	29 (A)	45 (B)	42 (C)	180
5	14 (E)	15 (A)	22 (B)	28 (C)	29 (D)	108
Totals	122	136	133	149	165	705

a. Conduct an analysis of variance for these data.
b. Is there sufficient evidence at the $\alpha = .05$ level of significance to indicate that a difference in mean yields for the five varieties of corn exists?

c. If rows were used to eliminate the water trend and columns were used to eliminate the fertility trend, test for significant row and column effects.

d. Based on your conclusions in part c., would a randomized block design, using rows as blocks, have been sufficient for this experiment?

e. Estimate the mean difference in yield between variety B and variety E with a 95% confidence interval.

2. In an investigation of the relative merits of four different methods of producing a chemical compound, a manufacturer, using four different suppliers of raw material, decided that a Latin Square design would eliminate the variability in quality from supplier to supplier, as well as variability from day to day. The following data were collected for y, the number of kilograms of compound produced:

		Suppliers			
Days	1	2	3	4	Totals
1	104 (B)	98 (D)	95 (C)	116 (A)	413
2	87 (D)	112 (B)	105 (A)	89 (C)	393
3	90 (C)	108 (A)	82 (D)	98 (B)	378
4	118 (A)	114 (C)	110 (B)	103 (D)	445
Totals	399	432	392	406	1629

a. Perform an analysis of variance for these data.

b. Test for a significant difference among methods of production at the .01 level of significance.

c. Estimate the mean difference in production between methods A and B with a 95% confidence interval.

d. Is there evidence of a difference among suppliers at the .05 level of significance?

e. Estimate the mean difference in production between suppliers 1 and 2 with a 95% confidence interval.

f. Was blocking on days effective?

13.11 Selecting the Sample Size

The quantity of information in a sample pertinent to a population parameter can be measured by the width of the confidence interval for the parameter. Since the width of the confidence interval for a single mean or the difference between two means is inversely proportional to the number of observations in a treatment mean, when a bound on the error of estimation is given, the sample sizes can be chosen large enough to achieve the required accuracy.

The selection of the sample size involves the following steps:

1. The experimenter must decide on the parameter or parameters of interest.
2. Second, he must specify a maximum bound on the error he is willing to tolerate.
3. An estimate of σ^2 must be available. This estimate may be obtained from a prior experiment, or roughly computed as ¼ of the expected range of the measurements to be taken.

4. The initial solution, which can be refined if so desired, involves the inequality

$$2\, \sigma_{\hat{\theta}} \leqslant B,$$

where 2 is used as an approximate value of $t_{.025}$ or $z_{.025}$ for 95% confidence.

Example 13.15
A completely randomized design is to be used in an experiment to compare the mean response time to a standard dosage of a drug using four different formulations. The experimenter would like to estimate the difference in mean response times correct to within 30 seconds with 95% accuracy. If the range of response times is expected to be about three minutes, how many observations should be included in each sample to achieve the required accuracy?

Solution
The 95% confidence interval estimator for $\mu_i - \mu_j$ is given as

$$(\bar{T}_i - \bar{T}_j) \pm t_{.025}\, s\, \sqrt{\frac{1}{n_i} + \frac{1}{n_j}}.$$

1. Since $t_{.025}$ depends upon the degrees of freedom available for estimating σ^2, we shall take $t_{.025} \approx$ _____.   **2**
2. When all populations have the same variance, the optimal solution is to take $n_i = n_j =$ _____.   **n**
3. Using the range to approximate σ, we have

$$s \approx \frac{R}{4} = \frac{\underline{\qquad}}{4} = \underline{\qquad}.$$ **3; .75**

4. For a maximum bound of 30 seconds = .5 minute, we need to solve

$$t_{.025}\, s\, \sqrt{\frac{1}{n_i} + \frac{1}{n_j}} \leqslant .5.$$

This is approximately

$$2(.75)\sqrt{\frac{2}{n}} \leqslant .5,$$

$$\left[\frac{2(.75)\sqrt{2}}{.5}\right]^2 \leqslant n,$$

or

$$n \geqslant \underline{\qquad}.$$ **18**

18

The experimenter should take equal samples of size _____ or more in each of the four groups to achieve the required accuracy.

Example 13.16
An experiment to compare the effect of three stimuli upon reaction times is to be run in a randomized block design using subjects as blocks. Previous experimentation in this area produced a standard deviation of 7.3 seconds. How many subjects should be included in the experiment if the experimenter wishes that the error in estimating the difference in two mean reaction times be less than or equal to four seconds with probability .95?

Solution
To estimate the difference in two treatment means in a randomized block design, we use

$$(\bar{T}_i - \bar{T}_j) \pm t_{.025} \, s \, \sqrt{\frac{1}{b} + \frac{1}{b}}.$$

7.3; 4; 2

When $s = $ _____ seconds, $B = $ _____ seconds and $t_{.025} \approx$ _____ we solve

$$2(7.3) \sqrt{\frac{2}{b}} \leqslant 4.$$

Solving for b, we find

$$b \geqslant \left[\frac{2(7.3)\sqrt{2}}{4} \right]^2,$$

26.65

$$b \geqslant \underline{\hspace{2cm}}.$$

27

Therefore, the experimenter should include at least _____ subjects in the experiment.

    The solutions to these sample size problems are only approximate because of the approximations used in arriving at the results, for example, $t_{.025} \approx 2$ or $s \approx R/4$. These solutions are very useful nonetheless, since they do give the experimenter the approximate *size* and therefore the approximate

cost

_____ of the experiment. For a further review of problems involving the selection of the sample size to achieve specified bounds of error in estimation, refer to Section 8.5 and reread your solutions to the examples given there.

EXERCISES

1. A large piece of cotton fabric was cut into 12 pieces and randomly partitioned into three groups of four. Three different chemicals designed to produce resistance to stain were applied to the units, one chemical for each group. A stain was applied (as uniformly as possible) over all

$n = 12$ units and the intensity of the stain measured in terms of light reflection.

a. What type of experimental design was employed?
b. Perform an analysis of variance and construct the *ANOVA* table for the following data:

	Chemical	
1	*2*	*3*
12	14	9
8	9	7
9	11	9
6	10	5

c. Do the data present sufficient evidence to indicate a difference in mean resistance to stain for the three chemicals?
d. Give a 95% confidence interval for the difference in means for chemicals 1 and 2.
e. Give a 90% confidence interval for the mean stain intensity for chemical 2.
f. Approximately how many observations per treatment would be required to estimate the difference in mean response for two chemicals correct to within 1.0?
g. Obtain *SSE* directly for the data of Exercise 1 by calculating the sums of squares of deviations within each of the three treatments and pooling. Compare with the value found using $SSE = $ Total $SS - SST$.

2. A substantial amount of variation was expected in the amount of stain applied to the experimental units of Exercise 1. It was decided that greater uniformity could be obtained by applying the stain three units at a time. A repetition of the experiment produced the following results:

		Chemical		
Application	*1*	*2*	*3*	Totals
1	12	15	9	36
2	9	13	9	31
3	7	12	7	26
4	10	15	9	34
Totals	38	55	34	127

a. Give the type of design.
b. Conduct an analysis of variance for the data.
c. Do the data provide sufficient evidence to indicate a difference between chemicals?
d. Give the formula for a $(1 - \alpha)$ 100% confidence interval for the difference in a pair of chemical means. Calculate a 95% confidence interval for $(\mu_2 - \mu_3)$.
e. Approximately how many blocks (applications) would be required to estimate $(\mu_1 - \mu_2)$ correct to within .5?
f. We noted that the chemist suspected an uneven distribution of stain when simultaneously distributed over the 12 pieces of cloth. Do the

data support this view? (That is, do the data present sufficient evidence to indicate a difference in mean response for applications?)

3. Twenty maladjusted children were randomly separated into four equal groups and subjected to three months of psychological treatment. A slightly different technique was employed for each group. At the end of the three-month period, progress was measured by a psychological test. The scores are shown below (one child in group 3 dropped out of the experiment).

	Group				
	1	2	3	4	
	112	111	140	101	
	92	129	121	116	
	124	102	130	105	
	89	136	106	126	
	97	99		119	
Totals	514	577	497	567	2155

a. Give the type of design which appears appropriate.
b. Conduct an analysis of variance for the data.
c. Do the data present sufficient evidence to indicate a difference in mean response on the test for the four techniques?
d. Find a 95% confidence interval for the difference in mean response on the test for groups 1 and 2.
e. How could one employ blocking to increase the information in this problem? Under what circumstances might a blocking design applied to this problem fail to achieve the objective of the experiment?

4. The Graduate Record Examination scores were recorded for students admitted to three different graduate programs in a university.

	Graduate Programs		
	1	2	3
	532	670	502
	601	590	607
	548	640	549
	619	710	524
	509		542
	627		
	690		

a. Do these data provide sufficient evidence to indicate a difference in mean level of achievement on the GRE for applicants admitted to the three programs?
b. Find a 90% confidence interval for the difference in mean GRE scores for programs 1 and 2.

5. In the investigation of four brands of typewriters, four typists were asked to use each of the typewriters for four specific periods of time. In addition to completing a questionnaire about the merits of each brand of type-writer, the number of words typed per minute was recorded for each

typist for a given brand and time period. A Latin Square design was used in order to eliminate the effect of the time periods and the typists' abilities.

		Time Periods		
Typists	1	2	3	4
1	68 (D)	52 (A)	62 (C)	56 (B)
2	59 (B)	72 (C)	53 (A)	56 (D)
3	70 (C)	71 (D)	55 (B)	49 (A)
4	43 (A)	52 (B)	50 (D)	64 (C)

a. Perform an analysis of variance for these data.
b. Is there a significant difference in performance (measured in typing speed) among the four brands of typewriters at the .05 level of significance?
c. Is there a significant difference in the performance of the four typists?
d. Would you conclude that blocking on time periods was necessary?
e. Construct a 95% confidence interval for the mean difference in words typed per minute between brands C and D.

6. In a study where the objective was to investigate methods of reducing fatigue among employees whose job involved a monotonous assembly procedure, twelve randomly selected employees were asked to perform their usual job under each of three trial conditions. As a measure of fatigue, the experimenter used the total length of time in minutes of assembly line stoppages during a four-hour period for each trial condition. The data follow.

		Conditions	
Employee	1	2	3
1	31	22	26
2	20	15	23
3	26	21	18
4	21	12	22
5	12	16	18
6	13	19	23
7	18	7	16
8	15	9	12
9	21	11	26
10	15	15	19
11	11	14	21
12	18	11	21

a. Perform an analysis of variance for these data, testing whether there is a significant difference among the mean stoppage times for the three conditions.
b. Is there a significant difference in mean stoppage times for the twelve employees? Was the blocking effective?
c. Estimate the difference in mean stoppage time for conditions 2 and 3 with 95% confidence.

Chapter 14

NONPARAMETRIC STATISTICS

14.1 Introduction

In earlier chapters we tested various hypotheses concerning populations in terms of their parameters. These tests represent a group of tests that are called _____ tests, since they specifically involve parameters such as means, variances, or proportions. In order to apply the techniques of Chapter 8, a large number of observations were required to assure the approximate _____ of the statistics employed in testing. In Chapters 9, 10, and 13, it was assumed that the sampled populations had _____ distributions. Further, if two or more populations were studied in the same experiment, it was necessary to assume that these populations had a common _____. In this chapter we shall be concerned with hypotheses that do not involve population parameters directly, but rather deal with the form of the distribution. The hypothesis that two distributions are identical versus the hypothesis that one distribution has typically larger values than the other are nonparametric statements of H_0 and H_a.

parametric

normality

normal

variance

Nonparametric tests are appropriate in many situations of interest where one or more of the following conditions exist:
1. Nonparametric methods can be used when the form of the distribution is unknown, so that descriptive parameters may be of little use.
2. Nonparametric techniques are particularly appropriate if the measurement scale is that of rank ordering.
3. If a response can be measured on a continuous scale, a nonparametric method may nevertheless be desirable because of its relative simplicity when compared to its parametric analogue.
4. Most parametric tests require that the sampled population satisfy certain assumptions. When an experimenter cannot reasonably expect that these assumptions are met, a nonparametric test would be a valid alternative.

The following hypotheses would be appropriate for nonparametric tests:
1. H_0: The population is normally distributed.
2. H_0: Populations I and II have the same distribution.
3. H_0: A sequence of observations exhibits the property of randomness.

less

Since these hypotheses are less specific than those required for parametric tests, we might expect a nonparametric test to be (more, less) efficient than a corresponding parametric test when all the conditions required for the use of the parametric test are met.

14.2 A Comparison of Statistical Tests

The conditions of an experiment are often such that two or more different tests would be valid for testing the hypotheses of interest. How could we compare the efficiencies of two such tests? Statisticians examine the power of a test and use power as a measure of efficiency. The power of a test is defined to be _____. If β is the probability that H_0 is accepted when H_a is true, then the complement of this event, $1 - \beta$, is the probability that H_0 is _____ when H_a is true. Since the object of a statistical test is to reject H_0 when it is _____, $1 - \beta$ represents the probability that the test will perform its designated task.

$1 - \beta$

rejected
false

One method of comparing two tests utilizing the same sample size and the same significance level (α) is to compare their powers for alternatives of concern to the experimenter. The most common method of comparing two tests is to find the relative efficiency of one test with respect to the other. Since the sample sizes represent a measure of the costs of the tests in question, we would choose the test requiring (fewer, more) sample observations to achieve the same level of significance (α) and the same power $(1 - \beta)$ as the (more, less) efficient test. If n_A and n_B denote the sample sizes required for tests A and B to achieve the same specified values of α and $1 - \beta$ for a specific alternative hypothesis, then the relative efficiency of test A with respect to test B is _____. If this ratio is greater than one, test A is said to be (more, less) efficient than B.

fewer

more

n_B/n_A
more

14.3 The Sign Test for Comparing Two Populations

signs

The sign test is based on the _____ of the observed differences. Thus in a paired-difference experiment, we may observe in each pair only whether the first element is larger than the second. If the first element is larger (smaller), we assign a plus (minus) sign to the difference. We will define the test statistic, y, to be the number of _____ signs observed.

plus

It is worth emphasizing that the sign test *does not* require a numerical measure of a response, but merely a statement of which of two responses within a matched pair is larger. Thus the sign test is a convenient and even necessary tool in many psychological investigations. If within a given pair it is impossible to tell which response is larger (a tie occurs), the pair is omitted. Thus if 20 differences are analyzed and two of them are impossible to classify as plus or minus, we shall base our inference on _____ (give number) differences.

18

Let p denote the probability that a difference selected at random from the population of differences would be given a plus sign. If the two population distributions are identical, the probability of a plus sign for a given pair would equal _____ . Then the null hypothesis, "the two populations are identical," could be stated in the form $H_0: p = 1/2$. The test statistic, y, will have a _____ distribution whether H_0 is true or not. If H_0 is true, then the number of trials, n, will be the number of pairs in which a difference can be detected and the probability of success (a plus sign) on a given trial will be _____ . If the alternative hypothesis is $H_a: p > 1/2$, then (large, small) values of y would be placed in the rejection region. If the alternative hypothesis is $H_a: p < 1/2$, then (large, small) values of y would be used in the rejection region. With $H_a: p \neq 1/2$, the rejection region would include both _____ and _____ values of y.

1/2

binomial

$p = 1/2$
large
small

large; small

Example 14.1

Thirty matched pairs of schizophrenic patients were used in an experiment to determine the effect of a certain drug on sociability. In eighteen of these pairs the patient receiving the drug was judged to be more sociable, while in five pairs it was not possible to detect a difference in sociability. Test whether this drug tends to increase sociability.

Solution

Let p denote the probability that, in a matched pair selected at random, the patient receiving the drug will be more sociable. Then, let y denote the number of pairs in which the drugged patient is more sociable.

1. The null hypothesis that the two populations are identical is stated as
   $H_0: p =$ _____ .
   The alternative hypothesis that the drugged patients will be more sociable
   can be written as $H_a:$ _____ .

   1/2

   $p > 1/2$

2. The test statistic is y, the number of responses in which the drugged patient was more sociable out of the $n = 25$ pairs in which a difference was detected. For this problem, $y =$ _____ .

   18

3. Using $y = 18, 19, \ldots , 25$ as a rejection region yields the value α
   = _____ . (Use the binomial tables in your text.) If $y = 17, 18, \ldots ,$
   25 is taken as the rejection region, $\alpha =$ _____ . Assuming that
   $\alpha = .05$ is a satisfactory significance level, we shall use the (first, second)
   rejection region.

   .022
   .054
   second

4. Since observed $y = 18$, we agree to reject H_0 and conclude that the drug tends to increase sociability among schizophrenics.

We observed in Chapter 7 of the text that the normal approximation to binomial probabilities is reasonably accurate when $p = 1/2$ even when n is as small as _____ . Thus, the normal distribution can ordinarily be used to approximate α and β for a given rejection region. Furthermore, when n is at least 25, the test can be based on the statistic

10

$$\frac{y - (n/2)}{.5 \sqrt{n}}$$

which will approximately the _____ _____ distribution when H_0 is true.

standard normal

Example 14.2

The productivity of 35 students was observed and measured both before and after the installation of new lighting in their classroom. The productivity of 21 of the 35 students was observed to have improved while the productivity of the others appeared to show no perceptible gain as a result of the new lighting. Test whether the new lighting was effective in increasing student productivity.

Solution

Let p denote the probability that one of the 35 students selected at random exhibits increased productivity after the installation of new lighting. This constitutes a paired-difference test where the productivity measures are paired on the students. Such pairing tends to block out student variations.

1. The null hypothesis is $H_0 : p$ _____. [= 1/2]
2. The appropriate one-sided alternative hypothesis is $H_a : p$ _____. [> 1/2]
3. If y denotes the number of students who show improved productivity after the installation of the new lighting, then y has a binomial distribution with mean $np = 35(1/2) =$ _____ [17.5] and variance $npq = 35(1/2)(1/2)$ = _____ [8.75]. Therefore, the test statistic can be taken to be

$$z = \frac{y - 17.5}{\sqrt{8.75}}.$$

4. Reject H_0 at the $\alpha = .05$ level of significance if the calculated value of z is greater than $z_{.05} =$ _____. [1.645]
5. Since $y = 21$,

$$z = \frac{21 - 17.5}{\sqrt{8.75}} = \underline{\qquad}.$$ [1.18]

Hence we (would, would not) reject H_0; the new lighting (has, has not) improved student productivity. [would not; has not]

14.4 The Mann-Whitney U Test for Comparing Two Population Distributions

If an experimenter has two independent (not related) random samples in which the measurement scale is at least rank ordering, he can test the hypothesis that the two underlying distributions are identical versus the alternative that they are not identical by using the Mann-Whitney U test.

If two independent random samples are drawn from the same population (this is the case if H_0 is true), then we really have one large sample of size $N = n_1 + n_2$. If all measurements were ranked from small (1) to large (N), and each observation from sample 1 replaced with an A, and each observation from sample 2 replaced with a B, we would expect to find the A's and B's randomly mixed in the ranking positions. If H_0 is false and the second sample comes from a population whose values tend to be larger than the first population, the B's would tend to occupy the _____ [higher] ranks. However, if the second sample comes from a population whose values tend to be smaller than the first, then the B's would appear in the _____ rank positions.

lower

A statistic that reflects the positions of the n_1 and n_2 sample values in the total ranking is the sum of the ranks occupied by the first sample or the sum of the ranks occupied by the second sample. This information is used directly by the Mann-Whitney U statistic which counts the number of times that an A observation precedes a B observation in the total ranking.

Example 14.3

Suppose that we have two samples each of size 5 with the following values:

$$\text{Sample } A: \quad 19 \quad 20 \quad 16 \quad 12 \quad 23$$
$$\text{Sample } B: \quad 17 \quad 21 \quad 22 \quad 25 \quad 18$$

1. Ranking all $N = 10$ observations, we have

A	A	B	B	A	A	B	B	A	B
12	16	17	18	19	20	21	22	23	25

2. For each value from sample B, count the number of A's that precede it in the total ranking.

For 17, the first B measurement in the ranking, the number of A measurements preceding it is $u_1 = 2$.

For 18, the second B measurement in the ranking, the number of A's preceding it is $u_2 = 2$. For 21, $u_3 =$ _____, for 22, $u_4 =$ _____, and for 25, $u_5 =$ _____.

4; 4
5

3. The sum

$$U = u_1 + u_2 + u_3 + u_4 + u_5 = \underline{\hspace{2cm}}$$

17

counts the number of times that an A precedes a B in the total ranking.

4. If the B's all occupy the _____ ranks, U will take its maximum value, indicating that the measurements in the B population tend to be larger than the A population. If all the B's occupy the _____ ranks, U will take its minimum value of 0, indicating that the measurements in the A population tend to be larger than the B population. Therefore, U is a statistic that will reflect the truth or falseness of H_0.

higher

lower

Although the value of U can always be calculated in the manner suggested above, it can also be simply calculated using the sum of the ranks assigned to the A observations. If n_1 is the number of A observations and n_2 is the number of B observations with T_A being the sum of the A ranks, then

$$U = n_1 n_2 + \tfrac{1}{2} n_1 (n_1 + 1) - T_A.$$

To verify the value $U = 17$ for Example 14.3, we note that
1. $n_1 = n_2 = 5$ and $\quad T_A = 1 + 2 + 5 + 6 + 9 = 23.$
2. Substituting,

23

15; 23

17

$$U = 5(5) + \tfrac{1}{2}(5)(6) - \underline{\hspace{2cm}}$$

$$= 25 + \underline{\hspace{2cm}} - \underline{\hspace{2cm}}$$

$$= 40 - 23$$

$$= \underline{\hspace{2cm}}.$$

When the alternative hypothesis is that the two populations differ in distribution, a two-tailed rejection region is appropriate. In that case, we calculate

$$U = n_1 n_2 + \tfrac{1}{2} n_1 (n_1 + 1) - T_A$$

and

$$U' = n_1 n_2 + \tfrac{1}{2} n_2 (n_2 + 1) - T_B.$$

large; small

small

Since U' counts the number of times that a B precedes an A, U will be small if U' is \underline{\hspace{2cm}} and U will be large if U' is \underline{\hspace{2cm}}. Hence, the smaller of the two is used as our sample value of U, and we would reject H_0 if U or U' is too \underline{\hspace{2cm}}.

The tabled values for $P[U \leqslant U_0]$ when n_1 and n_2 are 10 or less are given in Table 8 of your text. For a two-tailed test the tabulated probabilities must be doubled to find α, the level of significance.

Example 14.4
Before filling several new teaching positions at the high school level, the principal of the school formed a review board consisting of five teachers who were asked to interview the twelve applicants and rank them in order of merit. Seven of the twelve applicants held college degrees but had limited teaching experience. Of the remaining five applicants, all had college degrees and substantial experience. The review board's rankings are given below:

Limited Experience	Substantial Experience
4	1
6	2
7	3
9	5
10	8
11	
12	

Do these rankings indicate that the review board considers experience a prime factor in the selection of the best candidates?

Solution
1. In testing the null hypothesis that the underlying populations are identical versus the alternative hypothesis that the population consisting of applicants having substantial experience is better qualified (will receive low ranks), we require a \underline{\hspace{2cm}}-tailed test.

one

2. In deciding upon the test statistic and the rejection region, we must take care to note that the tables are given with $n_1 \leqslant n_2$. Hence we take $n_1 = 5$ and $n_2 = 7$ and identify the five applicants with substantial experience as A's and the remaining seven applicants as B's. If H_a is true, the A's will occupy the _____ ranks and U; the number of   lower
times that an A precedes a B in the ranking will be (small, large) while   large
U', the number of times that a B precedes an A in the ranking will be
(small, large).   small

3. Using U' as the test statistic, with $\alpha \approx .05$, $n_1 = 5$ and $n_2 = 7$, an appropriate rejection region would consist of the values $U \leqslant$ _____ with   7
$\alpha =$ _____ .   .0530

$$U' = n_1 n_2 + \frac{n_2 (n_2 + 1)}{2} - T_B$$

$$= (5)(7) + \frac{7(8)}{2} - \text{_____}$$ 59

$$= \text{_____} .$$ 4

4. Since the observed value of U falls in the rejection region, we (reject, do   reject
not reject) H_0 and conclude that the review board (does, does not)   does
consider the applicants with teaching experience to be more highly
qualified than those without.

When the sample sizes both exceed ten, Table 8 can no longer be used to
locate rejection regions for tests involving the Mann-Whitney U-statistic.
However, when the sample sizes exceed ten, the distribution of U can be
approximated by a _____ distribution with mean   normal

$$E(U) = \tfrac{1}{2} n_1 n_2$$

and variance

$$\sigma_U^2 = \frac{1}{12} n_1 n_2 (n_1 + n_2 + 1).$$

Therefore, as a test statistic we can use

$$z = \frac{U - E(U)}{\sigma_U}$$

with the appropriate one- or two-tailed rejection region expressed in terms of
z, the standard normal random variable.

Example 14.5

A manufacturer uses a large amount of a certain chemical. Since there are
just two suppliers of this chemical, the manufacturer wishes to test whether
the percentage of contaminants is the same for the two sources against the
alternative that there is a difference in the percentage of contaminants for
the two suppliers. Data from independent random samples are given below:

Supplier	*Contaminant Percentages*				
	.86	.69	.72	.65	1.13
A	.65	1.18	.45	1.41	.50
	1.04	.41			
	.55	.40	.22	.58	.16
B	.07	.09	.16	.26	.36
	.20	.15			

Solution

1. We combine the obtained contaminant percentages in a single ordered arrangement and identify each percentage by letter.

Percentage	.07	.09	.15	.16	.16	.20	.22	.26
Rank	1	2	3	4.5	4.5	6	7	8
Supplier	B	B	B	B	B	B	B	B
Percentage	.36	.40	.41	.45	.50	.55	.58	.65
Rank	9	10	11	12	13	14	15	16.5
Supplier	B	B	A	A	A	B	B	A
Percentage	.65	.69	.72	.86	1.04	1.13	1.18	1.41
Rank	16.5	18	19	20	21	22	23	24
Supplier	A	A	A	A	A	A	A	A

2. Since the sample sizes of $n_1 = 12$ and $n_2 = 12$ are beyond those given in Table 8, we can use the normal approximation to the distribution of U. The manufacturer, in asking whether there is a difference between the two
two- suppliers, has specified a ____________ tailed test. Therefore we would reject H_0 if U were either too large or too small. (For a two-tailed test using the normal approximation, we are at liberty to use either U or U' as the value of U to be tested.)
72; 300 Using $n_1 = n_2 = 12, E(U) =$ ________ and $\sigma_U^2 =$ ________.

$$U = n_1 n_2 + \tfrac{1}{2} n_1(n_1 + 1) - T_A$$

216 $= 144 + 78 -$ ________

6 $=$ ________,

while

$$U' = n_1 n_2 + \tfrac{1}{2} n_2(n_2 + 1) - T_B$$

84 $= 144 + 78 -$ ________

138 $=$ ________.

3. The rejection region in terms of $z = (U - E(U))/\sigma_U$ would be to reject H_0
1.96 if $|z| >$ ________. With $U = 6$,

$$z = \frac{6 - 72}{\sqrt{300}} = \frac{-66}{17.32} = \underline{\qquad}.$$

-3.81

4. Hence we would conclude that there (is, is not) a significant difference in contaminant percentages for the two suppliers.

   Had we used U' as the value of U, our result would have been

is

$$z = \frac{138 - 72}{\sqrt{300}} = \frac{66}{17.32} = \underline{\qquad}$$

3.81

and we would have arrived at the same conclusion.

   Use of the Mann-Whitney U-test eliminates the need for the restrictive assumptions of Student's t-test which requires that the samples be randomly drawn from _____ populations having _____ variances.

normal; equal

Self-Correcting Exercises 14A

1. An experiment was designed to compare the durabilities of two highway paints, paint A and paint B, under actual highway conditions. An A strip and a B strip were painted across a highway at each of 30 locations. At the end of the test period, the experimenter observed the following results: At 8 locations paint A showed the least wear, at 17 locations paint B showed the least wear, and at the other 5 locations the paint samples showed the same amount of wear. Can we conclude that paint B is more durable? (Use $\alpha = .05$.)

2. In a deprivation study to test the strength of two physiological drives, ten rats who were fed the same diet according to a feeding schedule were randomly divided into two groups of 5 rats. Group I was deprived of water for 18 hours and group II was deprived of food for 18 hours. At the end of this time, each rat was put into a maze having the appropriate reward at the exit and the time required to run the maze was recorded for each rat. The results follow, with time measured in seconds:

Water	Food
16.8	20.8
22.5	24.7
18.2	19.4
13.1	28.9
20.2	25.3

   Is there a difference in strength of these two drives as measured by the time required to find the incentive reward? Use the Mann-Whitney U-test with $\alpha \approx .05$.

3. Rootstock of varieties A and B was tested for resistance to nematode intrusion. An A and a B were planted side by side in each of ten widely separated locations. At the conclusion of the experiment all roots were brought into the laboratory for a nematode count. The results are recorded below.

Location	1	2	3	4	5	6	7	8	9	10
Variety A	463	268	871	730	474	432	538	305	173	592
Variety B	277	130	522	610	482	340	319	266	205	540

Can it be said that varieties A and B differ in their resistance to nematode intrusion? (Use a two-tailed sign test with $\alpha = .02$.)

4. The score on a certain psychological test, P, is used as an index of status frustration. The scale ranges from $P = 0$ (low frustration) to $P = 10$ (high frustration). This test was administered to independent random samples of seven men and eight women with the following results:

	Status-Frustration Score							
Women	6	10	3	8	8	7	9	
Men	3	5	2	0	3	1	0	4

Use the Mann-Whitney U-statistic with α as close to .05 as possible to test whether the distribution of status-frustration scores is the same for the two groups against the alternative that the status-frustration scores are higher among women.

14.5 The Wilcoxon Rank Sum Test for a Paired Experiment

sign

absolute values

smaller

absolute values
omitted
Mann-Whitney U

One previously discussed nonparametric test which may be used for a paired-difference experiment is the _____ test. While the sign test requires only the direction of the difference within each matched pair, a more efficient test is available if in addition the _____ _____ of the differences can be ranked in order of magnitude. The Wilcoxon rank sum test employs as a test statistic, T, the (smaller, larger) sum of ranks for differences of the same sign where the differences are ranked in order of their _____ _____. In calculating T, zero differences are _____ and ties in the absolute values of nonzero differences are treated in the same manner as prescribed for the _____-test. Critical values of T are given in Table 9.

Example 14.6
Twelve matched pairs of brain-damaged children were formed for an experiment to determine which of two forms of physical therapy is the more effective. One child was chosen at random from each pair and treated over a period of several months using therapy A, while the other child was treated during this period using therapy B. There was judged to be no difference within two of the matched pairs at the end of the treatment period. The results are summarized in the following table:

Pair	Difference Favorable to Treatment	Rank for the Absolute Value of the Difference
1	A	9
2	A	5
3	B	1.5
4	A	4
5	A	1.5
6	*	*
7	A	7
8	A	8
9	*	*
10	A	10
11	A	6
12	A	3

*zero difference

For a two-sided test with $\alpha = .05$, we should reject H_0: "treatments equally effective" when $T \leqslant$ _____. The sample value of T is _____.
Hence, we _____ H_0.
    It can be shown that the expected value and variance of T are:

8; 1.5
reject

$$E(T) = \frac{1}{4} n(n + 1)$$

$$\sigma_T^2 = \frac{1}{24} n(n + 1)(2n + 1)$$

where n is the number of _____ in the experiment. When n is at least 25 we may employ the test statistic

pairs

$$z = \frac{T - E(T)}{\sigma_T}$$

which will have approximately the _____ _____ distribution when H_0 is true.

standard normal

Example 14.7
A drug was developed for reducing the cholesterol level in heart patients. The cholesterol levels before and after drug treatment were obtained for a random sample of 25 heart patients with the following results:

	Cholesterol Level				Cholesterol Level	
Patient	Before	After		Patient	Before	After
1	257	243		13	364	343
2	222	217		14	210	217
3	177	174		15	263	243
4	258	260		16	214	198
5	294	295		17	392	388
6	244	236		18	370	357
7	390	383		19	310	299
8	247	233		20	255	258
9	409	410		21	281	276
10	214	216		22	294	295
11	217	210		23	257	227
12	340	335		24	227	231
				25	385	374

Test whether this drug has an effect on the cholesterol level of heart patients.

Solution
Differences, Before–After, arranged in order of their absolute values are shown below together with the corresponding ranks. Fill in the missing ranks.

Difference	Rank		Difference	Rank
- 1	2		7	14
- 1	2		- 7	14
- 1	2		7	14
- 2	4.5		8	16
- 2	4.5		11	_____
3	6.5		11	_____
- 3	6.5		13	19
- 4	8.5		14	_____
4	8.5		14	_____
5	11		16	
5	11		20	23
5	11		21	24
			30	25

Suppose the alternative hypothesis of interest to the experimenter is the statement, "the drug has the effect of reducing cholesterol levels in heart patients." Thus, the appropriate rejection region for $\alpha = .05$ is $z < -1.645$ where, in calculating z, we take T to be the smaller sum of ranks (the sum of ranks of the _____ differences).
When H_0 is true,

$$E(T) = \frac{1}{2} n(n + 1) = \underline{\hspace{2cm}}$$

and

$$\sigma_T^2 = \frac{1}{24} n(n + 1)(2n + 1) = \underline{\hspace{2cm}}.$$

17.5
17.5

20.5
20.5
22

negative

325

1381.25

Thus, we shall reject H_0 at the $\alpha = .05$ significance level if

$$z = \frac{T - 325}{\sqrt{1381.25}} < -1.645.$$

Summing the ranks of the negative differences, we obtain $T =$ _____ and hence, $z =$ _____ . Comparing z with its critical value, we _____ H_0 in favor of the alternative hypothesis that the drug has the effect of reducing cholesterol levels in heart patients.

	44
	-7.56
	reject

It is interesting to see what conclusion is obtained by using the sign test. Recall that y is equal to the number of positive differences and that the test statistic

$$z = \frac{y - n/2}{\sqrt{n}/2}$$

has approximately the _____ _____ distribution when n is greater than ten and $H_0: p = 1/2$ is true. With $\alpha = .05$ the rejection region for z is z _____ . But $y = 17$, so that $z =$ _____ . Thus, we obtain the same conclusion as before, though the sample value of the test statistic does not penetrate as deeply into the rejection region as when the Wilcoxon test was used. Since the Wilcoxon test makes fuller use of the information available in the experiment, we say that the Wilcoxon test is more _____ than the sign test.

	standard normal
	> 1.645; 1.8
	efficient

Self-Correcting Exercises 14B

1. The sign test is not as efficient as the Wilcoxon rank sum test for data of the type presented in Exercise 3, Self-Correcting Exercises 14A. Analyze the data of Exercise 3, Self-Correcting Exercises 14A, by using the two-tailed Wilcoxon test with $\alpha = .02$. Can it be said that varieties A and B differ in their resistance to nematode intrusion?

2. The sign test is sometimes used as a "quick and dirty" substitute for more powerful tests which require lengthy computations. The following differences were obtained in a paired-difference experiment: $-.93, .95, .52, -.26, -.75, .25, 1.08\ 1.47, .60, 1.20, -.65, -.15, 2.50, 1.22, .80, 1.27, 1.46, 3.05, -.43, 1.82, -.56, 1.08, -.16, 2.64.$
   Use the sign test with $\alpha = .05$ to test $H_0: \mu_d = 0$ against the one-sided alternative $H_a: \mu_d > 0$.

3. Refer to Exercise 2. Use the large sample Wilcoxon test with $\alpha = .05$ to test $H_0: \mu_d = 0$ against the alternative hypothesis $H_a: \mu_d > 0$. Compare (in efficiency and in computational requirements) the sign test and the Wilcoxon test as substitute tests in a paired-difference experiment.

14.6 The Runs Test: A Test for Randomness

The data for a runs test is obtained in the form of a sequence where each element is either a "success" (S) or a "failure" (F). A run is a maximal

like

5

nonrandomness

2

$C^{n_1+n_2}_{n_1}$
sample points

$C^{n_1+n_2}_{n_1}$

10

1.96

5; 7
.06
would not

sequence of _____ elements. The number of runs in the sequence, SSFSFFFSSS, is _____. A very small or very large number of runs in a sequence would indicate _____. If R is the number of runs in a sequence we shall reject H_0: "this sequence is random" if $R \leqslant K_1$ or if $R \geqslant K_2$ for suitably chosen values of K_1 and K_2. If there is at least one failure and at least one success, then the minimum value for R is _____.

If n_1 is the number of S elements and n_2 is the number of F elements in the sequence, then the total number of distinguishable arrangements of the sequence is _____. We take these distinguishable arrangements as equally likely _____ when H_0 is true. When H_0 is true, the probability that R will assume a specific value, y, is the number of distinguishable arrangements for which R is equal to y divided by _____. The probability distribution for R with $n_1 \leqslant n_2$ and both n_1 and n_2 less than or equal to 10 is provided in Table 10. If $n_1 \geqslant n_2$, simply interchange these symbols.

When n_1 and n_2 are both greater than _____ one may use the large-sample test statistic,

$$z = \frac{R - E(R)}{\sigma_R},$$

in which the expected value and variance of R are

$$E(R) = 1 + \frac{2n_1 n_2}{n_1 + n_2}$$

and

$$\sigma_R^2 = \frac{2n_1 n_2 (2n_1 n_2 - n_1 - n_2)}{(n_1 + n_2)^2 (n_1 + n_2 - 1)}.$$

The rejection region for a two-tailed test with $\alpha = .05$ is $|z| \geqslant$ _____.

Example 14.8

A salesman has contacted 12 customers on a certain day. Let S represent a sale and F a failure to make a sale. The sequential record for the day was: SSSFFFFSSSSF. Is there evidence of nonrandomness in this sequence?

Solution

$n_1 = $ _____ and $n_2 = $ _____. If we agree to reject when $R \leqslant 3$ and when $R \geqslant 10$, then $\alpha = $ _____. With this rejection region we (would, would not) reject the hypothesis of randomness.

Example 14.9
Refer to the previous example. A lower tail test could be justified in the
following situation. Suppose the district sales manager had reason to believe
that this particular salesman was unusually sensitive to success and failure.
Thus, a failure to sell seemed to reduce his confidence which in turn reduced
his selling effectiveness. The opposite effect seemed to be true when a sale
was consummated. If this theory were correct, the number of runs would
tend to be considerably (more, less) than if H_0 were true. Hence, an appro-   less
priate test would utilize the rejection region $R \leqslant 3$ with $\alpha =$ ___   .015
or the rejection region $R \leqslant 4$ with $\alpha =$ ___. If the latter rejection   .076
region were used, the district sales manager would ___ H_0 reject
and perhaps enroll his salesman in a Dale Carnegie school.

Example 14.10
A control chart is widely used in industry to provide a sequential record on
some measured characteristic. This chart has a central line representing the
process average. A measurement shall be classified as S if above this line and
F if below. Does the following sequence indicate a lack of randomness in the
distribution of this measured characteristic over time? *SSSFFFSSSSSSSFFF
SFFFFSSSSSF*

Solution
Though n_1 and n_2 are too large to allow use of Table 10, both n_1 and n_2 are
greater than ten. Hence, the z-statistic can be used in a test of randomness.
Since $n_1 =$ ___, $n_2 =$ ___,   15; 11

$$E(R) = 1 + \frac{2n_1 n_2}{n_1 + n_2} = \underline{\qquad}$$ 13.7

$$\sigma_R^2 = \frac{2n_1 n_2 (2n_1 n_2 - n_1 - n_2)}{(n_1 + n_2)^2 (n_1 + n_2 - 1)} = 5.94$$

$$\sigma_R = \underline{\qquad}.$$ 2.44

The test statistic is

$$z = \frac{R - 13.7}{2.44}.$$

With $\alpha = .05$, a two-sided test would reject when $|z| > 1.96$. The sample value
of R is ___ and hence the sample value of z is ___. The   8; -2.34
decision is to ___ H_0.   reject

Example 14.11
A runs test can be used to study Example 14.7. We shall use the label S for a
positive difference and F for a negative difference. The sequence of ordered
differences produces the arrangement: *FFFFFSFFSSSSSFSSSSSSSSSSSS*.
The number of runs is $R =$ ___. n_1 (the number of S elements)   6
= ___. n_2 (the number of F elements) = ___. Though   17; 8

n_2 is less than ten, we shall for illustrative purposes employ the large sample test statistic,

$$z = \frac{R - E(R)}{\sigma_R}$$

11.9

$$E(R) = 1 + \frac{2n_1 n_2}{n_1 + n_2} = \underline{\hspace{3cm}}$$

4.48

$$= \frac{2n_1 n_2 (2n_1 n_2 - n_1 - n_2)}{(n_1 + n_2)^2 (n_1 + n_2 - 1)} = \underline{\hspace{3cm}}.$$

To compare the runs test with the one-tailed Wilcoxon test, we shall reject

-1.645
-2.79; reject

H_0 at the level $\alpha = .05$ when $z < \underline{\hspace{2cm}}$. The sample value of z is $\underline{\hspace{2cm}}$, and hence we $\underline{\hspace{2cm}}$ H_0.

14.7 Rank Correlation Coefficient, r_s

The Spearman rank correlation coefficient, r_s, is a numerical measure of the association between two variables, y and x. As implied in the name of the

ranks

test statistic, r_s makes use of $\underline{\hspace{3cm}}$ and hence the exact value of numerical measurements on y and x need not be known. Conveniently r_s is

r
random

computed in exactly the same manner as $\underline{\hspace{2cm}}$, Chapter 10.

To determine whether variables y and x are related, we select a $\underline{\hspace{2cm}}$ sample of n experimental units (or items) from the population of interest. Each of the n items is ranked first according to the variable x and then

y
ranks

according to the variable $\underline{\hspace{2cm}}$. Thus, for each item in the experiment we obtain two $\underline{\hspace{2cm}}$. (Tied ranks are treated as in other parts of this chapter.) Let x_i and y_i denote the respective ranks assigned to item i. Then,

$$r_s = \frac{n \Sigma x_i y_i - (\Sigma x_i)(\Sigma y_i)}{\sqrt{[n \Sigma x_i^2 - (\Sigma x_i)^2][n \Sigma y_i^2 - (\Sigma y_i)^2]}}.$$

Example 14.12
An investigator wished to determine whether "leadership ability" is related to the amount of a certain hormone present in the blood. Six individuals were selected at random from the membership of the Junior Chamber of Commerce in a large city and ranked on the characteristic "leadership ability." A determination of hormone content for each individual was made from blood samples. The leadership ranks and hormone measurements are recorded below. Fill in the missing hormone ranks. Note that no difference in leadership ability could be detected for individuals 2 and 5.

Individual	Leadership Ability Rank (y_i)	Hormone Content	Hormone Rank (x_i)	
1	6	131	1	
2	3.5	174	_____	3
3	1	189	_____	5
4	2	200	6	
5	3.5	186	_____	4
6	5	156	_____	2

To calculate r_s we form an auxiliary table. Fill in the missing quantities.

Individual	y_i	y_i^2	x_i	x_i^2	$x_i y_i$	
1	6	_____	1	1	6	36
2	3.5	12.25	3	9	_____	10.5
3	1	1	5	_____	5	25
4	2	4	_____	36	12	6
5	_____	12.25	4	16	14	3.5
6	5	25	2	4	_____	10
Total	_____	90.5	21	91	57.5	21

Thus,

$$n \Sigma x_i y_i - (\Sigma x_i)(\Sigma y_i) = (\underline{})(57.5) - (21)(21) \qquad 6$$

$$= \underline{} \qquad -96$$

$$n \Sigma x_i^2 - (\Sigma x_i)^2 \quad = 6(\underline{}) - (21)^2 \qquad 91$$

$$= \underline{} \qquad 105$$

$$n \Sigma y_i^2 - (\Sigma y_i)^2 \quad = 6(90.5) - (\underline{})^2 \qquad 21$$

$$= \underline{}. \qquad 102$$

Finally,

$$r_s = \frac{-96}{\sqrt{(105)(\underline{})}} = -.93. \qquad 102$$

Thus high leadership ability (reflected in a low rank) seems to be associated with higher amounts of hormone.

The Spearman rank correlation coefficient may be employed as a test statistic to test an hypothesis of _____ between [no association] two characteristics. Critical values of r_s are given in Table 11. The tabulated quantities are values of r_0 such that $P[r_s > r_0] = .05, .025, .01$ or $.005$ as indicated. For a lower tail test, reject H_0: "no association between the two characteristics" when $r_s <$ _____. [$-r_0$]
    Two-tailed tests require doubling the stated values of α, and hence critical

0.10
0.05; 0.02; 0.01

values for two-tailed tests may be read from Table 11 if $\alpha =$ _____ ,
_____ , _____ , or _____ .

Example 14.13

Continuing Example 14.12, we may wish to test whether leadership ability
is associated with hormone level. If the experimenter had designed the
experiment with the objective of demonstrating that low leadership ranks
(high leadership abilities) are associated with high hormone levels, the

a lower
.829
-.829
does; do

appropriate test would be (a lower, an upper) tail test.
    For $\alpha = .05$ the critical value of r_s is $r_0 =$ _____ . Hence, we reject H_0
if $r_s <$ _____ . Since the sample value of r_s found in Example 14.12
(does, does not) fall in the rejection region we (do, do not) reject H_0.

Self-Correcting Exercises 14C

1. An automobile agency wished to study whether advertising has an effect
   on sales. The sales manager advertised only model A during the first week
   and only model B during the second week of the study. The sequential
   record of sales during the two-week period was $A, A, B, B, B, A, A, A, A,$
   $B, A, A, A, A, A, B, B, B, A, B, B, B, B, B, B, A, A$. If advertising increases
   the sales of the model advertised, the number of runs would tend to be
   less than the number expected in a random sequence. State the null
   hypothesis and test H_0 against the alternative that advertising increases
   sales of the model advertised. Use $\alpha = .05$.
2. Refer to Exercise 4, Self-Correcting Exercises 14A. Use a one-sided runs
   test, with α as close to .05 as possible, to test whether the distribution of
   status-frustration scores is the same in the two groups. The alternative
   hypothesis is that status-frustration scores are higher among the women.
3. An interviewer was asked to rank seven applicants as to their suitability
   for a given position. The same seven applicants took a written examination
   that was designed to rate an applicant's ability to function in the given
   position. The interviewer's ranking and the examination score for each
   applicant are given below.

Applicant	Interview Rank	Examination Score
1	4	49
2	7	42
3	5	58
4	3	50
5	6	33
6	2	65
7	1	67

   Calculate the value of Spearman's rank correlation for these data. Test for
   a significant negative rank correlation at the $\alpha = .05$ level of significance.
4. A manufacturing plant is considering building a subsidiary plant in another
   location. Nine plant sites are currently under consideration. After considering land and building costs, zoning regulations, available local work-
   force, and transportation facilities associated with each possible plant site,

two corporate executives have independently ranked the nine possible plant sites as follows:

Site	1	2	3	4	5	6	7	8	9
Executive 1	2	7	1	5	3	9	4	8	6
Executive 2	1	4	3	6	2	9	7	8	5

a. Calculate the Spearman's rank correlation between the two sets of rankings.
b. Is there reason to believe the two executives are in basic agreement regarding their evaluation of the nine plant sites? (Use $\alpha = .05$.)

EXERCISES

1. For each of the following tests, state whether the test would be used for related samples or for independent samples: sign test, Mann-Whitney U-test, Wilcoxon test, runs test.
2. About 1.2% of our combat forces in a certain area develop combat fatigue. To find identifying characteristics of men who are predisposed to this breakdown, the level of a certain adrenal chemical was measured in samples from two groups: men who had developed battle fatigue and men who had adjusted readily to combat conditions. The following determinations were recorded:

Battle Fatigue Group	23.35	21.08	22.36	20.24
	21.69	21.54	21.26	20.71
	20.00	23.40	21.43	21.54
	22.21			
Well-Adjusted Group	21.66	21.85	21.01	20.54
	20.19	19.26	21.16	19.97
	20.40	19.92	20.52	19.78
	21.15			

Use a large sample one-tailed Mann-Whitney test with α approximately equal to .05 to test whether the distributions of levels of this chemical are the same in the two groups against the alternative that the mean level is higher in the combat fatigue group.
3. An experiment was designed to determine whether exposure to cigarette smoke has an effect on the length of life of beagle dogs. Twenty beagles of the same age were used in the experiment. The animals were assigned at random to one of two groups. Ten of the dogs were subjected to conditions equivalent to smoking up to 12 cigarettes each day. The other ten acted as a control group. Recorded below is the number of days until death for the dogs in both groups. Since the experiment was concluded when the last of the experimental dogs died, an L is recorded for each of the dogs still living.

Experimental Group	Control Group
45	315
112	474
251	727
340	894
412	L
533	L
712	L
790	L
845	L
974	L

Use the Mann-Whitney U and $\alpha = .0526$ to obtain a one-tailed test of whether the experimental group has a shorter mean life than the control group.

4. Refer to Exercise 3. Use a one-tailed runs test with $\alpha = .051$ to test whether exposure to smoke shortens the expected life of beagle dogs. What do you surmise about the efficiency of the runs test relative to the Mann-Whitney test for detecting a difference in population means?

5. The value of r (defined in Chapter 10) for the following data is .636:

x	y
.05	1.08
.14	1.15
.24	1.27
.30	1.33
.47	1.41
.52	1.46
.57	1.54
.61	2.72
.67	4.01
.72	9.63

Calculate r_s for these data. What advantage of r_s was brought out in this example?

6. A ranking of the quarterbacks in the top eight teams of the National Football League was made by polling a number of professional football coaches and sports writers. This "true ranking" is shown below together with my ranking.

a. Calculate r_s.

b. Do the data provide evidence at the $\alpha = .05$ level of significance to indicate a positive correlation between my ranking and that of the experts?

Quarterback	A	B	C	D	E	F	G	H
True Ranking	1	2	3	4	5	6	7	8
My Ranking	3	1	4	5	2	8	6	7

7. Construction firms A and B are the only firms bidding for contracts in a certain area. Any cooperative arrangement between these firms would assure that any run of bids favorable to a given firm would be kept short

(and thus the number of runs would be high). Does the following sequence of winning bids indicate that the two firms are acting in collusion? Use $\alpha \leqslant .05$. *ABBABABAABABABBA*

Eight recent college graduates have interviewed for positions within the marketing department of a large industrial organization. The organization's Vice-President for Marketing and the Personnel Director rated each candidate independently on a 0–10 assumed interval scale. Their ratings are shown below. Use a two-tailed sign test to determine if the Vice-President and Personnel Manager differ in their evaluations of the eight candidates. (Use $\alpha \leqslant 0.10$.)

Candidate	1	2	3	4	5	6	7	8
Vice-President	3	7	6	9	7	4	3	8
Personnel Manager	2	4	5	5	9	1	6	6

Refer to Exercise 8 and analyze the data using the two-tailed Wilcoxon test with α as close as possible to the significance level used in Exercise 8. Explain any differences in conclusions arrived at using the Wilcoxon vs. the sign test. Which conclusion should we believe?

It is of interest to determine whether the efficiency of a certain machine operator is superior to the efficiency of another. To examine this question, the percentage of defective items produced daily by machine operators A and B are recorded over a period of time. More recorded data are available from operator A due to a recent illness experienced by operator B. The data are shown below.

		Percentage of Output Defective							
Operator A	3	2	7	6	5	5	3	7	4	6
Operator B	6	5	8	4	8	7	9	10		

The plant foreman is hesitant to use a t-test in the analysis as he realizes that distributions of percentages do not follow a normal distribution. Use the most powerful nonparametric test at your disposal to determine whether operator A produces a lower percentage defective than operator B. (Use $\alpha = .05$.)

SOLUTIONS TO
SELF–CORRECTING EXERCISES

Self-Correcting Exercises 2A

1. $f(4) = 2(4) + 3 = 11$.

2. $g(0) = 0^2 - 2 = -2$; $\ g(4) = 4^2 - 2 = 14$.

3. $h(0) = 0! = 1$; $\ h(4) = 4! = (4)(3)(2)(1) = 24$.

4. $g(0) + g(4) = 0! + 4! = 1 + 24 = 25$.

Set 2B

1. $\displaystyle\sum_{y=1}^{3} h(y) = h(1) + h(2) + h(3) = 2\left(\frac{1}{3}\right)^1 + 2\left(\frac{1}{3}\right)^2 + 2\left(\frac{1}{3}\right)^3$

$$= \frac{2}{3} + \frac{2}{9} + \frac{2}{27} = \frac{26}{27}.$$

2. $\displaystyle\sum_{x=1}^{4} (2x^2 - 5) = 2(1^2) - 5 + 2(2^2) - 5 + 2(3^2) - 5 + 2(4^2) - 5$

$$= 2 - 5 + 8 - 5 + 18 - 5 + 32 - 5 = 40.$$

3. $\displaystyle\sum_{i=1}^{10} x_i = (-1) + 2 + 1 + \ldots + (-2) = 3$.

4. $\displaystyle\sum_{i=1}^{10} x_i^2 = (-1)^2 + 2^2 + 1^2 + \ldots + (-2)^2 = 1 + 4 + 1 + \ldots + 4 = 97$.

5. $\displaystyle\sum_{i=1}^{10} x_i/10 = \frac{3}{10} = .3$.

Set 2C

1. $\displaystyle\sum_{x=1}^{10} 3 = 10(3) = 30$, using Theorem 2.1.

2. $\displaystyle\sum_{i=1}^{14} 4$ has 14 terms, while $\displaystyle\sum_{i=1}^{6} 4$ has 6 terms.

$$\sum_{i=7}^{14} 4 = \sum_{i=1}^{14} 4 - \sum_{i=1}^{6} 4 = 14(4) - 6(4) = 8(4) = 32.$$

3. a. $\displaystyle\sum_{i=3}^{8} (x_i - 4) = \sum_{i=3}^{8} x_i - 6(4) = 1 + 4 + (-3) + 1 + 6 - 24 = -15.$

 b. $\displaystyle\sum_{i=1}^{10} (x_i - 5)^2 = \sum_{i=1}^{10} x_i^2 - 2(5) \sum_{i=1}^{10} x_i + 10(5)^2$

$$= 97 - 10(3) + 250 = 317.$$

 See Example 2.13.

 c. $\displaystyle\sum_{i=1}^{5} (x_i^2 - x_i) = \sum_{i=1}^{5} x_i^2 - \sum_{i=1}^{5} x_i = 22 - 6 = 16.$

 d. $\displaystyle\sum_{i=1}^{10} x_i^2 - \left(\sum_{i=1}^{10} x_i\right)^2 \Bigg/ 10 = 97 - \frac{3^2}{10} = 97 - 0.9 = 96.1$

Set 3A

1. a. Range $= 59 - 18 = 41$.
 b.-c. Each student will obtain slightly different results. Dividing the range by 10 produces intervals of length slightly more than 4. A more convenient choice is to use 11 intervals of length 4, beginning at 17.5.

Class	Class Boundaries	Tally	f_i
1	17.5 – 21.5	1111	4
2	21.5 – 25.5	1111	4
3	25.5 – 29.5	⑴⑴⑴ 1	6
4	29.5 – 33.5	⑴⑴⑴ 11	7
5	33.5 – 37.5	⑴⑴⑴ 1	6
6	37.5 – 41.5	1111	4
7	41.5 – 45.5	1111	4
8	45.5 – 49.5	11	2
9	49.5 – 53.5	111	3
10	53.5 – 57.5	1	1
11	57.5 – 61.5	1	1

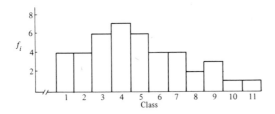

d. Dividing the range by 6, each interval must be of length 7.

Class	Class Boundaries	Tally	f_i
1	17.5 – 24.5	ⅢⅢ 11	7
2	24.5 – 31.5	ⅢⅢ ⅢⅢ	10
3	31.5 – 38.5	ⅢⅢ ⅢⅢ 1	11
4	38.5 – 45.5	ⅢⅢ 11	7
5	45.5 – 52.5	ⅢⅢ	5
6	52.5 – 59.5	11	2

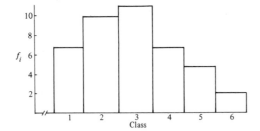

e. The second histogram is more informative, since it exhibits the "piling up" of the data in the middle classes. Using too many classes tends to "flatten out" the histogram, producing nearly equal frequencies in each class.

2. a. An extra column in the tabulation is used to calculate relative frequency.

Class	Class Boundaries	Tally	f_i	f_i/n
1	5.55 – 7.55	ⅢⅢ	5	5/32
2	7.55 – 9.55	ⅢⅢ	5	5/32
3	9.55 – 11.55	ⅢⅢ ⅢⅢ 11	12	12/32
4	11.55 – 13.55	ⅢⅢ	5	5/32
5	13.55 – 15.55	111	3	3/32
6	15.55 – 17.55	1	1	1/32
7	17.55 – 19.55	1	1	1/32

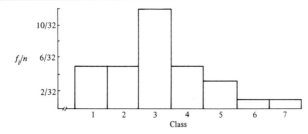

b. $\dfrac{1}{32} + \dfrac{1}{32} = \dfrac{2}{32}$ c. $\dfrac{5}{32} + \dfrac{5}{32} = \dfrac{10}{32}$ d. $\dfrac{12}{32} + \dfrac{5}{32} + \dfrac{3}{32} = \dfrac{20}{32}$

Set 3B

1. Arrange the set of data in order of ascending magnitude.

6	9	11	13	16
8	10	12	13	17
9	10	12	15	19

Median = 12 $\bar{y} = \dfrac{\displaystyle\sum_{i=1}^{n} y_i}{n} = \dfrac{180}{15} = 12$

2. Arrange the data in order of ascending magnitude.

y_i	$(y_i - \bar{y})$	$(y_i - \bar{y})^2$
0	-3	9
1	-2	4
2	-1	1
2	-1	1
2	-1	1
3	0	0
3	0	0
4	1	1
4	1	1
5	2	4
7	4	16
33	0	38

a. Median = 3 $\bar{y} = \dfrac{33}{11} = 3.$

b. Range = 7 - 0 = 7.

c. $s'^2 = \dfrac{\Sigma(y_i - \bar{y})^2}{n} = \dfrac{38}{11} = 3.4545$

$s' = \sqrt{3.4545} = 1.86.$

Set 3C

1. Display the data in a table as follows:

y_i	y_i^2
0	0
1	1
2	4
2	4
2	4
3	9
3	9
4	16
4	16
5	25
7	49
33	137

$$s'^2 = \frac{\Sigma y_i^2 - \dfrac{(\Sigma y_i)^2}{n}}{n} = \frac{137 - \dfrac{(33)^2}{11}}{11} = \frac{137 - 99}{11}$$

$$= \frac{38}{11} = 3.4545.$$

2. If $\bar{y}$ has been rounded off, then rounding error occurs each time $\bar{y}$ is subtracted from y_i in formula (a.) Hence, there are n possible rounding errors. If formula b. is used, only one rounding error occurs when $(\Sigma y_i)^2$ is divided by n. Hence, formula b. is less subject to rounding errors and results in a more accurate computation.

Set 3D

1. From Exercise 1, Self-Correcting Exercises 3C, we have $\Sigma y_i^2 = 137$ and $\Sigma y_i = 33$. Hence,

$$s^2 = \frac{\Sigma y_i^2 - \dfrac{(\Sigma y_i)^2}{n}}{n-1} = \frac{38}{10} = 3.8.$$

This is larger than $s'^2 = 3.4545.$

2. $\Sigma y_i = 29.7, \quad \Sigma y_i^2 = 129.19, \quad s^2 = \dfrac{\Sigma y_i^2 - \dfrac{(\Sigma y_i)^2}{n}}{n-1} = \dfrac{129.19 - \dfrac{(29.7)^2}{7}}{6}$

$$= \frac{129.19 - 126.0129}{6} = \frac{3.1771}{6} = .5295 \quad \text{and} \quad s = \sqrt{.5295} = .73.$$

3. $\Sigma y_i = 356, \quad \Sigma y_i^2 = 25362, \quad s^2 = \dfrac{25362 - \dfrac{356^2}{5}}{4} = \dfrac{25362 - 25347.2}{4}$

$$= \frac{14.8}{4} = 3.7 \quad \text{and} \quad \bar{y} = \frac{356}{5} = 71.2$$

Set 3E

1. a. If the value 165.0 were subtracted from each measurement, the transformed measurements would have values recorded to the nearest tenth, all lying between -5 and 5. If each measurement were next multiplied

by 10, the resulting measurements would be integers between -50 and 50. The most difficult arithmetic calculation would now be squaring a number between 0 and 50.

b. Let x_i be the coded measurement, and let y_i be the original measurement. Then $x_i = 10(y_i - 165) = 10y_i - 1650$. Using Theorem 3.3, $\bar{x} = 10\bar{y} - 1650$ and $s_x = 10s_y$.

2. The coded data are: 2, 5, 4, 1, 9, 6, 8, 5, 4, 5 with $\Sigma x_i = 49$, $\Sigma x_i^2 = 293$. Then

$$\bar{x} = \frac{49}{10} = 4.9$$

$$s_x^2 = \frac{293 - \dfrac{49^2}{10}}{9} = \frac{293 - 240.1}{9} = 5.8778.$$

Using Theorem 3.1, $\bar{y} = \bar{x} + 10 = 14.9$ and $s^2 = s_x^2 = 5.8778$.

3. a. The coding is $x_i = 100y_i - 50 = 100(y_i - \frac{1}{2})$, and the coded measurements are: 1, 3, -1, 1, 2, 3, 2, 0, 1, 1. Then $\Sigma x_i = 13$, $\Sigma x_i^2 = 31$.

b. $\bar{x} = \dfrac{13}{10} = 1.3$ and $s_x^2 = \dfrac{31 - \dfrac{(13)^2}{10}}{9} = \dfrac{14.1}{9} = 1.5667.$

Using Theorem 3.3, $\bar{y} = \dfrac{1}{100}(1.3) + \dfrac{1}{2} = .013 + .5 = .513$

$$s^2 = \frac{1}{(100)^2}(1.5667) = .00015667.$$

Set 4A

1. Denote the four good items as G_1, G_2, G_3, G_4, and the two defectives as D_1 and D_2.

a. $E_1: G_1G_2$ $E_4: G_1D_1$ $E_7: G_2G_4$ $E_{10}: G_3G_4$ $E_{13}: G_4D_1$

   $E_2: G_1G_3$ $E_5: G_1D_2$ $E_8: G_2D_1$ $E_{11}: G_3D_1$ $E_{14}: G_4D_2$

   $E_3: G_1G_4$ $E_6: G_2G_3$ $E_9: G_2D_2$ $E_{12}: G_3D_2$ $E_{15}: D_1D_2$

b. "At least one defective" implies one or two defectives, while "no more than one defective" implies zero or one defective.

$A: \{E_4, E_5, E_8, E_9, E_{11}, E_{12}, E_{13}, E_{14}, E_{15}\}.$

$B: \{E_4, E_5, E_8, E_9, E_{11}, E_{12}, E_{13}, E_{14}\}.$

$C: \{E_1, E_2, E_3, E_4, E_5, E_6, E_7, E_8, E_9, E_{10}, E_{11}, E_{12}, E_{13}, E_{14}\}.$

c. Each sample point is assigned equal probability; that is, $P(E_i) = 1/15$.

$P(A) = 9/15 = 3/5$ $P(B) = 8/15$ $P(C) = 14/15$.

2. a. E_1: *FFFF* E_5: *FFFM* E_9: *MFFM* E_{13}: *MFMM*

   E_2: *MFFF* E_6: *FFMM* E_{10}: *MFMF* E_{14}: *MMFM*

   E_3: *FMFF* E_7: *FMFM* E_{11}: *MMFF* E_{15}: *MMMF*

   E_4: *FFMF* E_8: *FMMF* E_{12}: *FMMM* E_{16}: *MMMM*

  b. A: $\{E_6, E_7, E_8, E_9, E_{10}, E_{11}\}$ B: $\{E_1\}$.

   C: $\{E_2, E_3, E_4, E_5, E_6, E_7, E_8, E_9, E_{10}, E_{11}, E_{12}, E_{13}, E_{14}, E_{15}, E_{16}\}$.

   $D = A \cup B$: $\{E_1, E_6, E_7, E_8, E_9, E_{10}, E_{11}\}$ $E = BC$: no sample points.

   $F = A \cup C$: same as C.

  c. Since each sample point is equally likely,

   $$P(A) = \frac{6}{16} = \frac{3}{8} \quad P(B) = \frac{1}{16} \quad P(C) = \frac{15}{16} \quad P(D) = \frac{7}{16} \quad P(E) = 0$$

   $$P(F) = \frac{15}{16} .$$

Set 4B

1. a. One camera is drawn at random from five, after which a second camera
     is chosen. Each is tested and found to be defective or nondefective.
  b. Since it is important whether the first or second camera is defective (see
     event D), there are 20 sample points to be listed. The first element in
     the pair denotes the first camera chosen.

   E_1: $G_1 G_2$ E_6: $G_2 D_1$ E_{11}: $G_2 G_1$ E_{16}: $D_1 G_2$

   E_2: $G_1 G_3$ E_7: $G_2 D_2$ E_{12}: $G_3 G_1$ E_{17}: $D_2 G_2$

   E_3: $G_1 D_1$ E_8: $G_3 D_1$ E_{13}: $D_1 G_1$ E_{18}: $D_1 G_3$

   E_4: $G_1 D_2$ E_9: $G_3 D_2$ E_{14}: $D_2 G_1$ E_{19}: $D_2 G_3$

   E_5: $G_2 G_3$ E_{10}: $D_1 D_2$ E_{15}: $G_3 G_2$ E_{20}: $D_2 D_1$

  c. A: $\{E_{10}, E_{20}\}$ B: $\{E_1, E_2, E_5, E_{11}, E_{12}, E_{15}\}$.

   C: $\{E_3, E_4, E_6, E_7, E_8, E_9, E_{10}, E_{13}, E_{14}, E_{16}, E_{17}, E_{18}, E_{19}, E_{20}\}$.

   D: $\{E_{10}, E_{13}, E_{14}, E_{16}, E_{17}, E_{18}, E_{19}, E_{20}\}$.

  d. $P(A) = \dfrac{2}{20} = \dfrac{1}{10} \quad P(B) = \dfrac{6}{20} = \dfrac{3}{10} \quad P(C) = \dfrac{14}{20} = \dfrac{7}{10} \quad P(D) = \dfrac{8}{20} = \dfrac{2}{5}$.

2. a. Since the pairs are ordered (see part b, Exercise 1), the number of
     sample points is $P_2^5 = (5)(4) = 20$.
  b. Event A: $P_2^2 = (2)(1) = 2$;   Event B: $P_2^3 = (3)(2) = 6$;
  c. Event C: $2P_1^2 P_1^3 + P_2^2 = 2(2)(3) + 2 = 14$ since the defective item may
     be chosen either first or second;

Event D: $P_1^2 \, P_1^3 + P_2^2 = 6 + 2 = 8$; since, once the first camera is found to be defective, the second can be either defective or good.

c. See Exercise 1d.

3. a.

E_1: HHH	E_3: NHN	E_5: HHN	E_7: NHH
E_2: HNN	E_4: NNH	E_6: HNH	E_8: NNN

Using the *mn* rule, the total should be $(2)(2)(2) = 8$.

b. $P(A) = \dfrac{3}{8}$ $P(B) = \dfrac{4}{8} = \dfrac{1}{2}$ $P(C) = \dfrac{1}{8}$ $P(D) = \dfrac{4}{8} = \dfrac{1}{2}$.

Set 4C

1. a. $A \cup C$: Same as C; AD: Same as A; CD: Same as D; $A \cup D$: Same as A.

b. $P(A \cup C) = \dfrac{7}{10}$; $P(AD) = \dfrac{1}{10}$; $P(CD) = \dfrac{2}{5}$; $P(A \cup D) = \dfrac{2}{5}$.

2. a. $D = A \cup B$; $E = BC$; $F = A \cup C$.

b. AB: no sample points; $B \cup C$: $\{E_1, E_2, \ldots, E_{16}\} = S$;

$AC \cup BC$: $\{E_6, E_7, E_8, E_9, E_{10}, E_{11}\}$; $\bar{C}$: $\{E_1\}$;

$\overline{AC}$: $\{E_1, E_2, \ldots E_5, E_{12}, E_{13}, \ldots E_{16}\}$.

c. $P(A \cup B) = P(A) + P(B) - P(AB) = \dfrac{3}{8} + \dfrac{1}{16} - 0 = \dfrac{7}{16}$

$P(\bar{C}) = 1 - P(C) = 1 - \dfrac{15}{16} = \dfrac{1}{16}$

$P(\overline{BC}) = 1 - P(BC) = 1 - 0 = 1$.

d. $P(A|C) = P(AC) / P(C) = \dfrac{6}{16} \Big/ \dfrac{15}{16} = \dfrac{6}{15}$ while $P(A) = \dfrac{3}{8}$.

A and C are dependent but are not mutually exclusive.

e. $P(B|C) = P(BC) / P(C) = 0 \Big/ \dfrac{15}{16} = 0$ while $P(B) = \dfrac{1}{16}$ and $P(BC) = 0$.

B and C are dependent and mutually exclusive.

3. a. $P(A) = P$ [the executive represents a small corporation] $= \dfrac{75}{200} = \dfrac{3}{8}$.

$P(F) = P$ [the executive favors gas rationing] $= \dfrac{15}{200} = \dfrac{3}{40}$.

$P(AF) = \dfrac{3}{200}$.

$P(A \cup G) = P$ [executive favors conversion or represents a small corporation or both] $= P(A) + P(G) - P(AG) = \dfrac{75 + 22 - 10}{200} = \dfrac{87}{200}$.

$P(AD) = P$ [executive represents a small corporation and favors car pooling] $= \dfrac{20}{200}$.

$$P(\bar{F}) = 1 - P(F) = 1 - \frac{3}{40} = \frac{37}{40}.$$

b. $P(A|F) = P(AF) / P(F) = \dfrac{3/200}{15/200} = \dfrac{3}{15} \qquad P(A|D) = \dfrac{20/200}{55/200} = \dfrac{20}{55}.$

Neither A and F nor A and D are mutually exclusive.
A and F and A and D are both dependent.

Set 4D

1. Define A: Company A shows an increase.
   B: Company B shows an increase.
   C: Company C shows an increase.
   It is given that $P(A) = .4$, $P(B) = 6$, $P(C) = 7$, and A, B, and C are independent events.

   a. $P(ABC) = P(A) P(B) P(C) = (.4) (.6) (.7) = .168$

   b. $P(\bar{A}\,\bar{B}\,\bar{C}) = P(\bar{A}) P(\bar{B}) P(\bar{C}) = [1 - P(A)]\ [1 - P(B)]\ [1 - P(C)] = (.6) (.4) (.3) = .072.$

   c. P [at least one shows profit] $= 1 - P$ [none show profit]
   $= 1 - P(\bar{A}\,\bar{B}\,\bar{C}) = 1 - .072 = .928.$

2. a. Let A be the event that the student can solve all five problems. The total number of sample points is $C_5^{10} = \dfrac{(10)\,(9)\,(8)\,(7)\,(6)}{(5)\,(4)\,(3)\,(2)\,(1)} = 252,$
   the number of possible ways to pick five questions from ten. The student will answer all five only if the professor selects all five from the six which he knows. This can happen in $C_5^6 = 6$ ways. Hence, $P(A) = \dfrac{6}{252}.$

   b. Define B: Student passes the exam.
   B_1: Student answers three correctly.
   B_2: Student answers four correctly.
   B_3: Student answers five correctly.

   Then $P(B) = P(B_1 \cup B_2 \cup B_3) = P(B_1) + P(B_2) + P(B_3)$

   $$= \frac{C_2^4\,C_3^6}{C_5^{10}} + \frac{C_1^4\,C_4^6}{C_5^{10}} + \frac{C_0^4\,C_5^6}{C_5^{10}}$$

   $$= \frac{6\,(20)}{252} + \frac{4\,(15)}{252} + \frac{6}{252} = \frac{186}{252}.$$

   c. $P(\text{fail}) = 1 - P(\text{pass}) = 1 - P(B) = 1 - \dfrac{186}{252} = \dfrac{66}{252}.$

Set 4E

1. Define the events F: A favorable seismic outcome results.
    O: Oil is actually present.
    $\bar{O}$: Oil is not present.

$$P(F|O) = .8; \quad P(F|\bar{O}) = .3; \quad P(O) = .5; \quad P(\bar{O}) = 1 - P(O) = .5.$$

$$P(O/F) = \frac{P(F|O)\, P(O)}{P(F|O)\, P(O) + P(F|\bar{O})P(\bar{O})} = \frac{(.8)\,(.5)}{(.8)\,(.5) + (.3)\,(.5)} = \frac{8}{11}.$$

2. Define the events D: A defective item is passed by an inspector.
    A_1: Inspector 1 inspected the item.
    A_2: Inspector 2 inspected the item.

$$P(A_1) = .6; \quad P(A_2) = .4; \quad P(D|A_1) = .01; \quad P(D|A_2) = .05.$$

$$P(A_1|D) = \frac{P(D|A_1)\, P(A_1)}{P(D|A_1)P(A_1) + P(D|A_2)\, P(A_2)} = \frac{.01(.6)}{.01(.6) + .05(.4)} =$$

$$\frac{.006}{.026} = .23.$$

Set 5A

1. The sample points are: (A, B), (A, C), (A, D), (B, C), (B, D), (C, D).
    If there is no discrimination in the selection, each sample point has
    probability $1/6$. Collecting information,

Sample Points	y	$p(y)$
(C, D)	0	1/6
$(A, C), (A, D), (B, C), (B, D)$	1	4/6
(A, B)	2	1/6

2. a. $p(0) = -2/10$ is not between 0 and 1.

    b. $\sum_y p(y) = 7/10 \neq 1$.

3. Denote the candidates as C_1, C_2, C_3, M_1, and M_2 where M_1 and M_2 are
    the two candidates with master's degrees. Since we are concerned only
    with the choice of the first and second ranked applicants, the 10 equally
    likely sample points are (C_1, C_2), (C_1, C_3), (C_1, M_1), (C_1, M_2), (C_2, C_3),
    (C_2, M_1), (C_2, M_2), (C_3, M_1), (C_3, M_2), (M_1, M_2). Collecting information,

Sample Points	y	$p(y)$
$(C_1, C_2), (C_1, C_3), (C_2, C_3)$	0	3/10
$(C_1, M_1), (C_1, M_2), (C_2, M_1)$ $(C_2, M_2), (C_3, M_1), (C_3, M_2)$	1	6/10
(M_1, M_2)	2	1/10

Set 5B

1. a. Since $\sum_y p(y) = 1$, $p(0) = 1 - \dfrac{6}{9} = \dfrac{3}{9}$.

b. $\mu = \sum_y y p(y) = -2\left(\dfrac{1}{9}\right) + (-1)\left(\dfrac{1}{9}\right) + 0\left(\dfrac{4}{9}\right) + 1\left(\dfrac{3}{9}\right)$

$$= \dfrac{-2}{9} - \dfrac{1}{9} + \dfrac{3}{9} = 0.$$

c. $\sigma^2 = E(y^2) - \mu^2 = (-2)^2\left(\dfrac{1}{9}\right) + (-1)^2\left(\dfrac{1}{9}\right) + 0^2\left(\dfrac{4}{9}\right) + 1^2\left(\dfrac{3}{9}\right) - 0^2$

$$= \dfrac{4}{9} + \dfrac{1}{9} + \dfrac{3}{9} = \dfrac{8}{9} = .8889$$

$\sigma = \sqrt{.8889} = .94$.

2. a. $\mu = E(y) = \sum_y y\, p(y) = 0(.1) + 1(.6) + 2(.2) + 3(.1) = .6 + .4 + .3 = 1.3$.

b. $\sigma^2 = E(y^2) - \mu^2 = 0^2(.1) + 1^2(.6) + 2^2(.2) + 3^2(.1) - (1.3)^2$

$$= .6 + .8 + .9 - 1.69 = 0.61.$$

3. $P[y \geqslant 2] = p(2) + p(3) = .2 + .1 = .3$.

4. Let y be the gain to the insurance company and let r be the premium charged by the company.

y	$p(y)$
r	.9900
$-15000 + r$	.0075
$-30000 + r$	.0025

In order to break even, $E(y) = 0$, or

$E(y) = \sum_y y\, p(y) = .99r + .0075\,(-15{,}000 + r)$

$\qquad\qquad + (-30{,}000 + r)(.0025) = 0$

$r - 112.50 - 75.00 = 0$

$r = \$187.50$.

Set 6A

1. Let y be the number of apartment dwellers who move within a year. Then $p = P$ [move within a year] $= .2$ and $n = 7$.

a. $P[y = 2] = C_2^7\,(.2)^2\,(.8)^5 = .27525$.

b. $P[y \leqslant 1] = C_0^7\,(.2)^0\,(.8)^7 + C_1^7\,(.2)^1\,(.8)^6 = .209715 + .367002 = .576717$.

2. Let y be the number of letters delivered within 4 days. Then $p = P$ [letter delivered within 4 days] $= .7$ and $n = 20$.

a. $P[y \geqslant 15] = 1 - P[y \leqslant 14] = 1 - .584 = .416$.

b. $P[y \geqslant 10] = 1 - P[y \leqslant 9] = 1 - .017 = .983$. Notice that if 10 or fewer letters arrive later than 4 days then $20 - 10 = 10$ or more will arrive within 4 days.

3. Use Table 1, Appendix II, indexing $n = 15, p = \dfrac{1}{2}$.

a. $P[y = 4] = P[y \leqslant 4] - P[y \leqslant 3] = .059 - .018 = .041$.

b. $P[y \geq 4] = 1 - P[y \leq 3] = 1 - .018 = .982$.

4. Let y be the number of fines given a student in twenty days, so that $p = P$ [fine] $= .10$. Using Table 1 with $n = 20$, $p = .1$,

a. $P[y = 0] = .122$.

b. $P[y \leq 4] = .957$

Set 6B

1. Let y be the number of accepted applicants taking a place in the freshman class, so that $p = .9$ and $n = 1360$; $\mu = np = 1360(.9) = 1224$; $\sigma = \sqrt{npq} = \sqrt{122.4} = 11.06$. The freshman class limits are $\mu \pm 2\sigma = 1224 \pm 22.12$ or 1201.88 to 1246.12. Choosing the largest interval with integer endpoints within these limits, the class will be between 1202 and 1246 in size.

2. Let y be the number of minority members on a list of 80, so that $p = .20$ and $n = 80$. Then $\mu = 80(.2) = 16$ and $\sigma = \sqrt{npq} = \sqrt{12.8} = 3.58$. The limits are $\mu \pm 2\sigma = 16 \pm 7.16$ or 8.84 to 23.16. We expect to see between 9 and 23 minority group members on the jury lists.

3. y = number watching the T.V. program.

$p = .4$.

$n = 400$.

$$\mu = 400(.4) = 160; \quad \sigma^2 = 400(.4)(.6) = 96; \quad \sigma = \sqrt{96} = 9.80.$$

Calculate $\mu \pm 2\sigma = 160 \pm 2(9.8) = 160 \pm 19.6$ or 140.4 to 179.6. Since we would expect the number watching the show to be between 141 and 179 with probability .95, it is highly unlikely that only 96 people would have watched the show *if* the 40% claim is correct. It is more likely that the percentage of viewers for this particular show is less than 40%.

Set 6C

1. a. P (acceptance) $= C_0^4 \, p^0 \, q^4 + C_1^4 \, p^1 \, q^3$ for various values of p. When $p = 0$, P (acceptance) $= 1$; $p = .3$, P (acceptance) $= (.7)^4 + 4(.3)(.7)^3$

$$= .2401 + .4116$$

$$= .6517.$$

When $p = 1$, P (acceptance) $= 0$. The graph follows the procedures used in Example 6.14 and is omitted here.

b. Keep $n = 4$, take $a > 1$; keep $a = 1$, take $n < 4$.

2. a.

p	0	.1	.3	.5	1.0
$n = 10$, $a = 1$	1	.736	.149	.011	0
$n = 25$, $a = 1$	1	.271	.002	.000	0

b.–c. If the student will graph the 3 O.C. curves as given in part a and Exercise 1, he will see that increasing n has the effect of decreasing the probability of acceptance.

3. a.

p	0	.1	.3	.5	1.0
$n = 25$, $a = 3$	1	.764	.033	.000	0
$n = 25$, $a = 5$	1	.967	.193	.002	0

b.–c. Increasing a has the effect of increasing the probability of acceptance.

Set 6D

1. Let y be the number of times subject B agreed with subject A so that $p = P$ [B agrees with A] and $n = 20$. The hypothesis to be tested is

$$H_0: p = \frac{1}{2} \quad \text{vs.} \quad H_a: p > \frac{1}{2}.$$

Large values of the test statistic y would favor rejection of H_0 in favor of H_a. Hence, we seek a rejection region of the form $y \geq a$ so that

$$\alpha = P \left[y \geq a | p = \frac{1}{2} \right] \leq .05.$$

From Table 1, with $n = 20$ and $p = .5$, the rejection region is $y \geq 15$ with $\alpha = 1 - .979 = .021$. Since the observed value of y ($y = 16$) falls in the rejection region, H_0 is rejected. There is sufficient evidence to conclude that there is an influence due to suggestion.

2. y = number understanding size concept.
p = probability of grasping size concept.
$n = 25$.

$$H_0: p = .80; \quad H_a: p > .80.$$

Large values of y suggest H_a is true, and the rejection region must be of the form $y \geq a$, with

$$\alpha = P [y \geq a | p = .05] \leq .05$$

$$1 - P [y \leq a - 1] \leq .05$$

$$P [y \leq a - 1] \geq .95.$$

From Table 1, $a - 1 = 23$ or $a = 24$. The rejection region is $y \geq 24$ with $\alpha = 1 - .973 = .027$. The observed value ($y = 22$) does not fall in the rejection region; H_0 cannot be rejected. There is insufficient evidence to doubt the 80% figure. Note that we cannot "accept H_0" unless β, the probability of a type II error, is assessed for meaningful alternative values of $p > .8$.

3. y = number of sales of new color. $H_0: p = .4$.
$p = P$ [sale is of new color]. $H_a: p < .4$.
$n = 15$.
From Table 1, reject H_0 if $y \leq 2$ with $\alpha = .027$. The observed value is $y = 4$; do not reject H_0. There is no reason to doubt the 40% figure.

Set 7A

Note: the student should illustrate each problem with a diagram and list all pertinent information before attempting the solution. Diagrams are omitted in order to conserve space.

1. a. $P [z > 2.1] = .5000 - A (2.1) = .5000 - .4821 = .0179.$

 b. $P [z < -1.2] = .5000 - A (1.2) = .5000 - .3849 = .1151.$

 c. $P [.5 < z < 1.5] = P [0 < z < 1.5] - P [0 < z < .5]$

$$= A (1.5) - A (.5) = .4332 - .1915 = .2417.$$

d. $P[-2.75 < z < -1.70] = A(2.75) - A(1.7) = .4970 - .4554 = .0416.$

e. $P[-1.96 < z < 1.96] = A(1.96) + A(1.96) = 2(.4750) = .95.$

f. $P[z > 1.645] = .5000 - A(1.645) = .5000 - .4500 = .05.$

Linear interpolation was used in part f. That is, since the value $z = 1.645$ is halfway between two tabled values, $z = 1.64$ and $z = 1.65$, the appropriate area is taken to be halfway between the two tabled areas, $A(1.64) = .4495$ and $A(1.65) = .4505$. As a general rule, values of z will be rounded to two decimal places, except for this particular example, which will occur frequently in our calculations.

2. a. We know that $P[z > z_0] = .10$, or $.5000 - A(z_0) = .10$, which implies that $A(z_0) = .4000$.

The value of z_0 which satisfies this equation is $z_0 = 1.28$, so that $P[z > 1.28] = .10.$

b. $P[z < z_0] = .01$ so that $.5000 - A(z_0) = .01$ and $A(z_0) = .4900$. The value of z_0 which satisfies this equation is $z_0 = -2.33$ so that $P[z < -2.33] = .01$. The student who draws a diagram will see that z_0 must be negative, since it must be in the left-hand portion of the curve.

c. $P[-z_0 < z < z_0] = A(z_0) + A(z_0) = .95$ so that $A(z_0) = .4750$. The necessary value of z_0 is $z_0 = 1.96$ and $P[-1.96 < z < 1.96] = .95.$

d. $P[-z_0 < z < z_0] = 2A(z_0) = .99$ so that $A(z_0) = .4950$. The necessary value of z_0 is $z_0 = 2.58$ and $P[-2.58 < z < 2.58] = .99.$

3. The random variable of interest has a standard normal distribution and hence may be denoted as z.

a. $P[z > 1] = .5000 - A(1) = .5000 - .3413 = .1587.$

b. $P[z > 1.5] = .5000 - A(1.5) = .5000 - .4332 = .0668.$

c. $P[-1 < z < -.5] = A(1) - A(.5) = .3413 - .1915 = .1498.$

d. The problem is to find a value of z, say z_0, such that $P[-z_0 < z < z_0] = .95.$

This was done in 2 c. and $z_0 = 1.96$. Hence, 95% of the billing errors will be between $-1.96 and $1.96.

e. Undercharges imply negative errors. Hence, the problem is to find z_0 such that $P[z < z_0] = .05.$

That is, $.5000 - A(z_0) = .05$ or $A(z_0) = .4500$. The value of z_0 is $z_0 = -1.645$ (see 1(f)) and hence, 5% of the undercharges will be at least $1.65.

Set 7B

1. We have $\mu = 10$, $\sigma = \sqrt{2.25} = 1.5$.

a. $P[y > 8.5] = P\left[\dfrac{y - \mu}{\sigma} > \dfrac{8.5 - 10}{1.5}\right] = P[z > -1]$

$\qquad\qquad = .5000 + A(1) = .5000 + .3413 = .8413.$

b. $P[y < 12] = P\left[z < \dfrac{12 - 10}{1.5}\right] = P[z < 1.33] = .5000 + A(1.33)$

$$= .5000 + .4082 = .9082$$

c. $P[9.25 < y < 11.25] = P\left[\dfrac{9.25 - 10}{1.5} < z < \dfrac{11.25 - 10}{1.5}\right]$

$$= P[-.5 < z < .83] = .1915 + .2967 = .4882.$$

d. $P[7.5 < y < 9.2] = P[-1.67 < z < -.53] = .4525 - .2019 = .2506.$

e. $P[12.25 < y < 13.25] = P[1.5 < z < 2.17] = .4850 - .4332 = .0518.$

2. The random variable of interest is y, the length of life for a standard household lightbulb. It is normally distributed with $\mu = 250$ and $\sigma = \sqrt{2500} = 50$.

a. $P[y > 300] = P\left[z > \dfrac{300 - 250}{50}\right] = P[z > 1] = .5000 - .3413 = .1587.$

b. $P[190 < y < 270] = P[-1.2 < z < .4] = .3849 + .1554 = .5403.$

c. $P[y < 260] = P[z < .2] = .5000 + .0793 = .5793.$

d. It is necessary to find a value of y, say y_0, such that $P[y > y_0] = .90.$

Now, $P[y > y_0] = P\left[\dfrac{y - \mu}{\sigma} > \dfrac{y_0 - 250}{50}\right] = .90$, so that

$P\left[z > \dfrac{y_0 - 250}{50}\right] = .5 + A\left(\dfrac{y_0 - 250}{50}\right) = .90$, or $A\left(\dfrac{y_0 - 250}{50}\right) = .40.$

By looking at a diagram, the student will notice that the value satisfying this equation must be negative. From Table 3, this value, $\dfrac{y_0 - 250}{50}$, is

$\dfrac{y_0 - 250}{50} = -1.28$ or $y_0 = -1.28 \,(50) + 250 = 186.$ Ninety percent of the bulbs have a useful life in excess of 186 hours.

e. Similar to d. It is necessary to find y_0 such that $P[y < y_0] = .95.$ Now,

$P[y < y_0] = P\left[z < \dfrac{y_0 - 250}{50}\right] = .95$

$.5000 + A\left(\dfrac{y_0 - 250}{50.}\right) = .95$

$A\left(\dfrac{y_0 - 250}{50}\right) = .45.$

Hence, $\dfrac{y_0 - 250}{50} = 1.645$ or $y_0 = 332.25.$ That is, 95% of all bulbs will burn out before 332.25 hours.

3. The random variable is y, scores on a trade school entrance examination, and has a normal distribution with $\mu = 50$ and $\sigma = 5$.

a. $P[y > 60] = P\left[z > \dfrac{60 - 50}{5}\right] = P[z > 2] = .5000 - .4772 = .0228.$

b. $P[y < 45] = P[z < -1] = .5000 - .3413 = .1587$.

c. $P[35 < y < 65] = P[-3 < z < 3] = .4987 + .4987 = .9974$.

d. It is necessary to find y_0 such that $P[y < y_0] = .95$. As in 2 e.,

$$A\left(\frac{y_0 - 50}{5}\right) = .45.$$

$$\frac{y_0 - 50}{5} = 1.645 \quad \text{or} \quad y_0 = 58.225.$$

Set 7C

1. a. Using Table 1 with $n = 20$, $p = .7$, $P[10 \leqslant y \leqslant 16] = P[y \leqslant 16]$ − $P[y \leqslant 9] = .893 - .017 = .876$.
   b. The probabilities associated with the values $y = 10, 11, \ldots 16$ are needed; hence, the area of interest is the area to the right of 9.5 and to the left of 16.5. Further, $\mu = np = 14$, $\sigma^2 = npq = 4.2$.

$$P[10 \leqslant y \leqslant 16] \approx P[9.5 < y < 16.5]$$

$$= P\left[\frac{9.5 - 14}{\sqrt{4.2}} < z < \frac{16.5 - 14}{\sqrt{4.2}}\right]$$

$$= P[-2.20 < z < 1.22] = .4861 + .3888 = .8749.$$

2. Let $p = P$ [income is less than 12,000] = ½, since 12,000 is the median income. Also, $n = 100$, $\mu = np = 50$, $\sigma^2 = npq = 25$.

$$P[y \leqslant 37] \approx P[y < 37.5] = P\left[z < \frac{37.5 - 50}{5}\right]$$

$$= P[z < -2.5] = .5000 - .4938 = .0062.$$

The observed event is highly unlikely under the assumption that $12,000 is the median income. The $12,000 figure does not seem reasonable.

3. y = number of white seeds.

$p = P$ [white seed] = .4. $\mu = np = 40$.

$n = 100$. $\sigma^2 = npq = 24$ $\sigma = 4.90$.

a. $P[y \leqslant 50] \approx P[y < 50.5] = P\left[z < \frac{50.5 - 40}{4.9}\right] = P[z < 2.14]$

$$= .5 + .4848 = .9838.$$

b. $P[y \leqslant 35] \approx P[y \leqslant 35.5] = P[z < -.92] = .5 - .3212 = .1788$.

c. $P[25 \leqslant y \leqslant 45] \approx P[24.5 < y < 45.5] = P[-3.16 < z < 1.12]$

$$= .5 + .3686 = .8686.$$

4. $\mu \pm 2\sigma = 40 \pm 2(4.9) = 40 \pm 9.8 \quad \text{or} \quad 31 \text{ to } 49$.

Set 8A

1. $\bar{y} = 67.5$ with approximate bound on error $2 \dfrac{s}{\sqrt{n}} = 2 \dfrac{8.2}{\sqrt{93}} = 1.70$.

2. $\hat{p}_1 - \hat{p}_2 = \dfrac{y_1}{n_1} - \dfrac{y_2}{n_2} = \dfrac{31}{204} - \dfrac{41}{191} = .15 - .21 = -.06$.

Approximate bound on error: $2 \sqrt{\dfrac{\hat{p}_1 \hat{q}_1}{n_1} + \dfrac{\hat{p}_2 \hat{q}_2}{n_2}} = 2 \sqrt{.000625 + .000869}$

$$= 2\,(.039) = .078.$$

3. $\hat{p} = \dfrac{y}{n} = \dfrac{8}{50} = .16$ with approximate bound on error $2 \sqrt{\dfrac{\hat{p}\hat{q}}{n}} = 2 \sqrt{\dfrac{.16\,(.84)}{50}}$

$$= 2\,(.0518) = .1036.$$

4. $\bar{y}_1 - \bar{y}_2 = 150.5 - 160.2 = -9.7$ with approximate bound on error

$$2 \sqrt{\dfrac{s_1^2}{n_1} + \dfrac{s_2^2}{n_2}} = 2 \sqrt{\dfrac{23.72}{35} + \dfrac{36.37}{35}} = 2\sqrt{1.7169} = 2.62.$$

Set 8B

1. $\hat{p} \pm z_{\alpha/2} \sqrt{\dfrac{\hat{p}\hat{q}}{n}}$ where $\hat{p} = \dfrac{30}{65} = .46$; $.46 \pm 2.33 \sqrt{\dfrac{.46\,(.54)}{65}}$;

$.46 \pm 2.33\,(.0618)$; $.46 \pm .14$ or $.32$ to $.60$.

2. $\hat{p}_1 = \dfrac{50}{100} = .50$; $\hat{p}_2 = \dfrac{60}{200} = .30$

$(\hat{p}_1 - \hat{p}_2) \pm 1.96 \sqrt{\dfrac{\hat{p}_1 \hat{q}_1}{n_1} + \dfrac{\hat{p}_2 \hat{q}_2}{n_2}}$

$(.50 - .30) \pm 1.96 \sqrt{\dfrac{.5\,(.5)}{100} + \dfrac{.3\,(.7)}{200}}$

$.2 \pm 1.96\,(.0596)$; $.2 \pm .12$ or $.08$ to $.32$.

3. a. $(\bar{y}_1 - \bar{y}_2) \pm z_{\alpha/2} \sqrt{\dfrac{s_1^2}{n_1} + \dfrac{s_2^2}{n_2}}$

$(12520 - 11210) \pm 2.58 \sqrt{\dfrac{(1510)^2}{90} + \dfrac{(950)^2}{60}}$

$1310 \pm 2.58\,(200.938)$

1310 ± 518.42 or 791.58 to 1828.42.

b. If the two schools belonged to populations having the same mean annual income, then $\mu_1 = \mu_2$, or $\mu_1 - \mu_2 = 0$. This value of $\mu_1 - \mu_2$ does not fall in the confidence interval obtained above. Hence, it is unlikely that the two schools belong to populations having the same mean annual income.

4. $\bar{y} \pm z_{\alpha/2} \dfrac{s}{\sqrt{n}}$; $22 \pm 1.645 \dfrac{4}{\sqrt{39}}$; $22 \pm \dfrac{6.58}{6.245}$; 22 ± 1.05 or 20.95 to 23.05.

Set 8C

1. The estimator of μ is $\bar{y}$, with standard deviation $\sigma/\sqrt{n}$. Hence, solve

$$2\dfrac{\sigma}{\sqrt{n}} = B, \quad 2\dfrac{8}{\sqrt{n}} = 3, \quad \sqrt{n} = \dfrac{16}{3}, \quad n = 28.4.$$

The experimenter should obtain $n = 29$ measurements.

2. For each additive, the range of the measurements is 80, so that

$$\sigma_1 = \sigma_2 \approx \dfrac{\text{Range}}{4} = 20.$$ Assuming equal sample sizes are acceptable, solve

$$2\sqrt{\dfrac{\sigma_1^2}{n} + \dfrac{\sigma_2^2}{n}} = 10; \quad 2\sqrt{\dfrac{2(20)^2}{n}} = 10; \quad \sqrt{n} = \dfrac{\sqrt{800}}{5}; \quad n = 32.$$

Hence, 32 subjects should be included in each group.

3. Maximum variation occurs when $p_1 = p_2 = .5$. Again assuming equal sample sizes, solve

$$2\sqrt{\dfrac{p_1 q_1}{n} + \dfrac{p_2 q_2}{n}} = .01; \quad 2\sqrt{\dfrac{2\,(.5)\,(.5)}{n}} = .01; \quad \sqrt{n} = \dfrac{2\sqrt{.5}}{.01};$$

$n = 20{,}000$.

4. Maximum variation occurs when $p = .5$. Since 90% accuracy is involved, solve

$$1.645\sqrt{\dfrac{pq}{n}} = .05; \quad 1.645\sqrt{\dfrac{(.5)\,(.5)}{n}} = .05; \quad \sqrt{n} = \dfrac{1.645\,(.5)}{.05} = 16.45;$$

$n = 270.6$; 271 patient records should be sampled.

Set 8D

1. $H_0 : \mu_1 - \mu_2 = 0$; $H_a : \mu_1 - \mu_2 > 0$; Test statistic: $z = \dfrac{(\bar{y}_1 - \bar{y}_2) - 0}{\sqrt{\dfrac{s_1^2}{n_1} + \dfrac{s_2^2}{n_2}}}$.

With $\alpha = .05$, reject H_0 if $z > 1.645$. Calculate

$$z = \dfrac{720 - 693}{\sqrt{\dfrac{104}{50} + \dfrac{85}{50}}} = \dfrac{27}{1.94} = 13.92.$$

Reject H_0. There is a significant difference in the mean scores. Men score higher on the average than women.

2. $H_0: \mu = 3.35$; $H_a: \mu \neq 3.35$; Test statistic: $z = \dfrac{\bar{y} - \mu_0}{s/\sqrt{n}}$.

With $\alpha = .05$, reject H_0 if $|z| > 1.96$. Calculate

$$\bar{y} = \frac{1264.40}{400} = 3.161; \quad s^2 = \frac{4970.3282 - 3996.7684}{399} = 2.4400;$$

$$s = \sqrt{2.44} = 1.56; \quad z = \frac{\bar{y} - \mu_0}{s/\sqrt{n}} = \frac{3.161 - 3.35}{1.56/20} = -2.42$$

Reject H_0. Mean revenue is different from \$3.35.

3. $H_0: p = \dfrac{1}{2}$; $H_a: p \neq \dfrac{1}{2}$; Test statistic: $z = \dfrac{\hat{p} - p_0}{\sqrt{\dfrac{p_0 q_0}{n}}}$

With $\alpha = .10$, reject H_0 if $|z| > 1.645$. Since $\hat{p} = \dfrac{480}{900} = .533$,

$$z = \frac{.533 - .5}{\sqrt{\dfrac{.5(.5)}{900}}} = \frac{.033}{.5/30} = 1.98. \text{ Reject } H_0. \text{ Half the cases are not male.}$$

4. $H_0: p_1 - p_2 = 0$; $H_a: p_1 - p_2 \neq 0$; Test statistic: $z = \dfrac{(\hat{p}_1 - \hat{p}_2) - 0}{\sqrt{\hat{p}\hat{q}\left(\dfrac{1}{n_1} + \dfrac{1}{n_2}\right)}}$

Note that if $p_1 = p_2$ as proposed under H_0, the best estimate of this common value of p is

$$\hat{p} = \frac{y_1 + y_2}{n_1 + n_2} = \frac{16 + 6}{100 + 50} = .15.$$

With $\alpha = .05$, reject H_0 if $|z| > 1.96$. Calculate

$$z = \frac{\dfrac{16}{100} - \dfrac{6}{50}}{\sqrt{.15(.85)\left(\dfrac{1}{100} + \dfrac{1}{50}\right)}} = \frac{.04}{.0618} = .65.$$

Do not reject H_0. There is insufficient evidence to detect a difference in the performances of the machines.

Set 9A

1. $H_0: \mu = 10$; $H_a: \mu < 10$; Test statistic: $t = \dfrac{\bar{y} - \mu}{s/\sqrt{n}}$

With $\alpha = .01$ and $n - 1 = 9$ degrees of freedom, reject H_0 if $t < -t_{.01} = -2.821$. Calculate

$$t = \frac{9.4 - 10}{1.8/\sqrt{10}} = \frac{-.6}{.57} = -1.05.$$

Do not reject H_0. There is insufficient evidence to claim that the mean weight is less than 10 ounces.

$$\bar{y} \pm t_{.01} \frac{s}{\sqrt{n}}; \quad 9.4 \pm 2.821 \frac{1.8}{\sqrt{10}}; \quad 9.4 \pm 1.6 \text{ or } 7.8 \text{ to } 11.0.$$

2. $H_0: \mu = 35$; $H_a: \mu > 35$; Test statistic: $t = \dfrac{\bar{y} - \mu}{s/\sqrt{n}}$.

With $\alpha = .05$, and $n - 1 = 19$ degrees of freedom, reject H_0 if $t > t_{.05} = 1.729$. Calculate

$$t = \frac{42 - 35}{6.2/\sqrt{20}} = \frac{7}{1.386} = 5.05.$$

Reject H_0. The mean riding time is greater than 35 minutes.

3. $\bar{y} \pm t_{.025} \dfrac{s}{\sqrt{n}}$; $\quad 42 \pm 2.093 \, (1.386)$; $\quad 42 \pm 2.90$ or 39.1 to 44.9.

Set 9B

1. See Section 9.4, paragraph 1.

2. $H_0: \mu_1 - \mu_2 = 0$; $H_a: \mu_1 - \mu_2 \neq 0$; Test statistic: $t = \dfrac{(\bar{y}_1 - \bar{y}_2) - D_0}{s\sqrt{\dfrac{1}{n_1} + \dfrac{1}{n_2}}}$.

With $\alpha = .05$ and $n_1 + n_2 - 2 = 16 + 9 - 2 = 23$ degrees of freedom, reject H_0 if $|t| > t_{.025} = 2.069$. Calculate

$$s^2 = \frac{(n_1 - 1) s_1^2 + (n_2 - 1) s_2^2}{n_1 + n_2 - 2} = \frac{15 \, (2.8)^2 + 8 \, (5.3)^2}{23} = \frac{117.6 + 224.72}{23}$$

$$= 14.8835 \quad \text{and} \quad s = 3.86.$$

$$t = \frac{5.2 - 8.7}{3.86 \sqrt{\dfrac{1}{16} + \dfrac{1}{9}}} = \frac{-3.5}{3.86 \, (.42)} = -2.16.$$

Reject H_0. The mean distance to the health center differs for the two groups.

3. $(\bar{y}_1 - \bar{y}_2) \pm t_{.025} \, s \sqrt{\dfrac{1}{n_1} + \dfrac{1}{n_2}}$; $\quad -3.5 \pm 2.069 \, (1.62)$; -3.5 ± 3.35;

or -6.85 to -0.15 miles.

4. $H_0: \mu_1 - \mu_2 = 0$; $H_a: \mu_1 - \mu_2 \neq 0$; Test statistic: $t = \dfrac{(\bar{y}_1 - \bar{y}_2) - D_0}{s\sqrt{\dfrac{1}{n_1} + \dfrac{1}{n_2}}}$.

With $\alpha = .05$ and $n_1 + n_2 - 2 = 34$ degrees of freedom, reject H_0 if $|t| > t_{.025} = 1.96$. Calculate

$$s^2 = \frac{(n_1 - 1)s_1^2 + (n_2 - 1)s_2^2}{n_1 + n_2 - 2} = \frac{17\,(23.2) + 17\,(19.8)}{34} = 21.5.$$

$$t = \frac{81.7 - 77.2}{\sqrt{21.5\left(\frac{2}{18}\right)}} = \frac{4.5}{1.5456} = 2.91.$$

Reject H_0. There is a difference in mean scores for the two methods.

Set 9C

1. $H_0: \mu_d = \mu_P - \mu_H = 0$; $H_a: \mu_d = \mu_P - \mu_H < 0$; Test statistic: $t = \dfrac{\bar{d} - \mu_d}{s_d/\sqrt{n}}$.

With $\alpha = .05$ and $n - 1 = 10 - 1 = 9$ degrees of freedom, reject H_0 if $t < -t_{.05} = -1.833$. The 10 differences and the calculation of the test statistic are given below.

d_i	d_i^2
-5	25
-2	4
4	16
-3	9
-3	9
1	1
-1	1
0	0
-5	25
1	1
-13	91

$$\bar{d} = \frac{-13}{10} = -1.3 \quad s_d^2 = \frac{91 - \dfrac{(-13)^2}{10}}{9} = \frac{74.1}{9} = 8.2333.$$

$$s_d = \sqrt{8.2333} = 2.869$$

$$t = \frac{-1.3 - 0}{2.869/\sqrt{10}} = -1.43.$$

Do not reject H_0.

2.

d_i	d_i^2
4.6	21.16
1.8	3.24
-1.0	1.00
2.2	4.84
1.0	1.00
1.2	1.44
2.6	6.76
2.8	7.84
2.0	4.00
3.0	9.00
20.2	60.28

$$\bar{d} = \frac{20.2}{10} = 2.02 \quad s_d^2 = \frac{60.28 - 40.804}{9} = \frac{19.476}{9}$$

$$= 2.164.$$

$$s_d = 1.47.$$

a. $\bar{d} \pm t_{.025}\dfrac{s_d}{\sqrt{n}}$, $2.02 \pm 2.262\dfrac{1.47}{\sqrt{10}}$,

  2.02 ± 1.05 or $.97 < \mu_d < 3.07$.

b. $H_0: \mu_P - \mu_H = 0$; $H_a: \mu_P - \mu_H > 0$;

  Test statistic: $t = \dfrac{\bar{d} - \mu_d}{s_d/\sqrt{n}}$.

With $\alpha = .05$ and $n - 1 = 9$ degrees of freedom, reject H_0 if $t > t_{.05} = 1.833$. Calculate

$$t = \frac{2.02}{.465} = 4.34.$$

Reject H_0. Per-unit scale increases production.

Set 9D

1. $H_0 : \sigma = 3$ $(\sigma^2 = 9)$; $H_a : \sigma < 3$ $(\sigma^2 < 9)$; Test statistic: $\chi^2 = \dfrac{(n-1)s^2}{\sigma_0^2}$.

With $\alpha = .05$ and $n - 1 = 4$ degrees of freedom, reject H_0 if $\chi^2 < \chi_{.95}^2 = .710721$. Calculate

$$s^2 = \frac{\Sigma y_i^2 - \dfrac{(\Sigma y_i)^2}{n}}{n-1} = \frac{1930.5 - 1905.152}{4} = 6.337.$$

$$\chi^2 = \frac{25.348}{9} = 2.816. \quad \text{Do not reject } H_0.$$

2. $H_0 : \sigma^2 = 100$; $H_a : \sigma^2 < 100$; Test statistic: $\chi^2 = \dfrac{(n-1)s^2}{\sigma_0^2}$.

With $\alpha = .05$ and $n - 1 = 29$ degrees of freedom, reject H_0 if $\chi^2 < \chi_{.95}^2 = 17.7083$. Calculate

$$\chi^2 = \frac{29(8.9)^2}{100} = 22.97.$$

Do not reject H_0.

3. a. $H_0 : \sigma = 5$; $H_a : \sigma > 5$; Test statistic: $\chi^2 = \dfrac{(n-1)s^2}{\sigma_0^2}$.

With $\alpha = .05$, reject H_0 if $\chi^2 > 42.5569$.

Since $\chi^2 = \dfrac{29(7.3)^2}{25} = 61.81$, reject H_0. The new technique is less sensitive.

b. $\dfrac{(n-1)s^2}{\chi_U^2} < \sigma^2 < \dfrac{(n-1)s^2}{\chi_L^2}$, $\dfrac{29(7.3)^2}{45.7222} < \sigma^2 < \dfrac{29(7.3)^2}{16.0471}$,

$33.80 < \sigma^2 < 96.30$, $5.81 < \sigma < 9.81$.

Set 9E

1. $H_0 : \sigma_1^2 = \sigma_2^2$; $H_a : \sigma_1^2 \neq \sigma_2^2$; Test statistic: $F = \dfrac{s_1^2}{s_2^2}$ where population 1 is the population of distances for people wanting the center closed. With

$\alpha = .02$, and $v_1 = 8$, $v_2 = 15$, reject H_0 if $F > 4.00$. Calculate

$$F = \frac{(5.3)^2}{(2.8)^2} = \frac{28.09}{7.84} = 3.58.$$

Do not reject H_0. Assumption has been met.

2. H_0: $\sigma_1^2 = \sigma_2^2$; H_a: $\sigma_1^2 \neq \sigma_2^2$; Test statistic: $F = \dfrac{s_1^2}{s_2^2}$ where population 1 is the population of thresholds for females. With $\alpha = .10$, $v_1 = 12$, $v_2 = 9$, reject H_0 if $F > 3.07$. Calculate

$$F = \frac{26.9}{11.3} = 2.38.$$

Do not reject H_0. The two groups exhibit the same basic variation.

Set 10A

1. a.

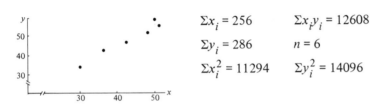

$\Sigma x_i = 256$ $\Sigma x_i y_i = 12608$

$\Sigma y_i = 286$ $n = 6$

$\Sigma x_i^2 = 11294$ $\Sigma y_i^2 = 14096$

The trend appears to be linear.

b. $SS_{xy} = 12608 - \dfrac{(256)(286)}{6} = 405.3333$

$SS_x = 11294 - \dfrac{(256)^2}{6} = 371.3333$

$\hat{\beta}_1 = \dfrac{SS_{xy}}{SS_x} = \dfrac{405.3333}{371.3333} = 1.09.$

$\hat{\beta}_0 = \dfrac{286}{6} - 1.09 \left(\dfrac{256}{6}\right) = 47.6667 - 46.5067 = 1.16$

c. $\hat{y} = 1.16 + 1.09\,(50) = 55.66$ or   5566 students.

2. a.

$\Sigma x_i = 31.6$ $\Sigma x_i y_i = 624.6$ $SS_{xy} = 624.6 - \dfrac{(31.6)(135)}{7} = 15.1714$

$\Sigma y_i = 135$ $n = 7$

$\Sigma x_i^2 = 149.82$ $\Sigma y_i^2 = 2645$ $SS_x = 149.82 - \dfrac{(316)^2}{7} = 7.1686$

$\hat{\beta}_1 = \dfrac{SS_{xy}}{SS_x} = \dfrac{15.1714}{7.1686}$

$\hat{\beta}_0 = 19.29 - 2.12\,(4.51) = 9.73.$

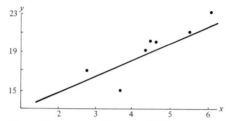

$\hat{y} = 9.73 + 2.12\,x.$

Predictor appears adequate.

3. a. $\Sigma x_i = 96$ $\Sigma x_i y_i = 1799$ $SS_{xy} = 1799 - \dfrac{(96)(135)}{7} = -52.4286$

   $\Sigma y_i = 135$ $n = 7$

   $\Sigma x_i^2 = 1402$ $\Sigma y_i^2 = 2645$ $SS_x = 1402 - \dfrac{(96)^2}{7} = 85.4286$

   $\hat{\beta}_1 = \dfrac{SS_{xy}}{SS_x} = \dfrac{-52.4286}{85.4286} = -0.61,$

   $\hat{\beta}_0 = 19.29 - (-.61)(13.71) = 27.65.$

   $y = 27.65 - .61_x.$

   b.

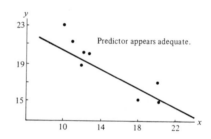

Predictor appears adequate.

Set 10B

1. $SSE = SS_y - \hat{\beta}_1\, SS_{xy} = SS_y - \dfrac{(SS_{xy})^2}{SS_x}$

   $= 14096 - \dfrac{(286)^2}{6} - \dfrac{(405.3333)^2}{371.3333}$

   $= 463.33 - 442.45 = 20.88.$

   Note that of $\hat{\beta}_1\, SS_{xy} = \dfrac{(SS_{xy})^2}{SS_x}$ has been used for the sake of accuracy.

   $s^2 = \dfrac{20.88}{4} = 5.22;\quad s = \sqrt{5.22} = 2.28.$

   a. $H_0: \beta_1 = 0.\quad H_a: \beta_1 \neq 0.$

   Reject H_0 if $|t| > t_{.025,4} = 2.776.$

   Test statistic: $t = \dfrac{\hat{\beta}_1 - \beta_1}{s/\sqrt{SS_x}} = \dfrac{1.09}{\sqrt{\dfrac{5.22}{371.33}}} = \dfrac{1.09}{.12} = 9.08.$ Reject $H_0.$

b. $\hat{\beta}_1 \pm t_{.025} \dfrac{s}{\sqrt{SS_x}} = 1.09 \pm 2.776\,(.12) = 1.09 \pm .33$

   or $.76 < \beta_1 < 1.42.$

2. $SSE = 2645 - \dfrac{1}{7}(135)^2 - \dfrac{(15.1714)^2}{7.1686} = 41.43 - 32.11 = 9.32.$

$s^2 = \dfrac{9.32}{5} = 1.86; \quad s = \sqrt{1.86} = 1.36.$

$H_0: \beta_1 = 0. \quad H_a: \beta_1 \neq 0.$

Rejection region: With 5 degrees of freedom and $\alpha = .05$, reject H_0 if $|t| > t_{.025} = 2.571.$

Test statistic: $t = \dfrac{\hat{\beta}_1 - 0}{s/\sqrt{SS_x}} = \dfrac{2.12}{\sqrt{\dfrac{1.86}{7.17}}} = \dfrac{2.12}{.51} = 4.16.$ Reject $H_0.$

3. $SSE = 2645 - \dfrac{1}{7}(135)^2 - \dfrac{(-52.4286)^2}{85.4286}$

   $= 41.43 - 32.18 = 9.25.$

$s^2 = \dfrac{9.25}{5} = 1.85; \quad s = 1.36.$

$H_0: \beta_1 = 0. \quad H_a: \beta_1 \neq 0. \quad$ Reject H_0 if $|t| > t_{.025,5} = 2.571.$

$t = \dfrac{-.61 - 0}{\sqrt{\dfrac{1.85}{85.43}}} = \dfrac{-.61}{.15} = -4.067.$ Reject $H_0.$

Set 10C

1. $H_0: E(y|x = 0) = 0. \quad H_a: E(y|x = 0) \neq 0.$

Test statistic: $t = \dfrac{\hat{y} - E_0}{s\sqrt{\dfrac{1}{n} + \dfrac{(x_0 - \bar{x})^2}{SS_x}}}$

Rejection region: With $\alpha = .05$, reject H_0 if $|t| > t_{.025} = 2.776.$ Calculate:

   $\hat{y} = 1.16 + 1.09\,(0) = 1.16$

   $t = \dfrac{1.16 - 0}{\sqrt{5.22\left[\dfrac{1}{6} + \dfrac{(42.67)^2}{371.33}\right]}} = \dfrac{1.16}{\sqrt{26.47}} = \dfrac{1.16}{5.14} = .226.$

Do not reject $H_0.$ Note that $\hat{y} = 1.16 + 1.09\,x$ does not pass through the origin, even though we could not reject the hypothesis, $H_0: E(y|x = 0) = \beta_0 = 0$. We could not reject H_0 because of the small number of observations and the large variation.

2. $(\hat{y}|x = 4.5) = 9.73 + 2.12 (4.5) = 9.73 + 9.54 = 19.27$

$$V(\hat{y}|x) = 1.87 \left[\frac{1}{7} + \frac{(4.50 - 4.51)^2}{7.17} \right]$$

$$= 1.87 (.14) = .2618.$$

The 90% confidence interval is $\hat{y} \pm t_{.05} \sqrt{V(y|x)}$

$$= 19.27 \pm 2.015 \sqrt{.2618}$$

$$= 19.27 \pm 1.03$$

or $18.24 < E(y|x = 4.5) < 20.30$.

Since $x = 250$ is outside the limits for the observed x, one should not predict for that value.

3. $(\hat{y}|x = 12) = 27.65 - .61 (12) = 20.33, \quad V(\hat{y}|x = 12)$

$$= 1.85 \left[\frac{1}{7} + \frac{(12 - 13.71)^2}{85.43} \right] = 1.85 (.18) = .33.$$ The 90% confidence interval

is $20.33 \pm 2.015 \sqrt{.33} = 20.33 \pm 2.015 (.57) = 20.33 \pm 1.15$ or

$19.18 < E(y|x = 12) < 21.48$. Note that this interval predicts a slightly higher expected yield.

4. $(\hat{y}|x = 40) = 1.16 + 1.09 (40) = 44.76; V(\hat{y}|x) = 5.22 \left[\frac{1}{6} + \frac{(40 - 42.67)^2}{371.33} \right]$

$$= 5.22 (.19) = .99.$$ The 95% confidence interval is

$44.76 \pm 2.776 \sqrt{.99} = 44.76 \pm 2.776$ or $41.98 < E(y|x = 40) < 47.54$.

Enrollment will be between 4,198 and 4,754 with 95% confidence.

Set 10D

1. a. $r = \dfrac{SS_{xy}}{\sqrt{SS_x \, SS_y}} = \dfrac{15.1714}{\sqrt{(7.1686)(41.43)}} = \dfrac{15.1714}{17.2335} = .88.$

b. $r^2 = (.88)^2 = .7744.$ Total variation is reduced by 77.44% by using number of cotton bolls to aid in prediction.

2. a. $r = \dfrac{-52.4286}{\sqrt{(85.4286)(41.43)}} = \dfrac{-52.4286}{59.4921} = -.88.$

Total variation is reduced by 77.44% by using number of damaging insects to aid in prediction.

b. The predictors are equally effective.

3.

x_1 (Bolls)	x_2 (Insects)
5.5	11
2.8	20
4.7	13
4.3	12
3.7	18
6.1	10
4.5	12

$\Sigma x_1 = 31.6$ $\quad$ $\Sigma x_2 = 96$

$\Sigma x_1^2 = 149.82$ $\quad$ $\Sigma x_2^2 = 1402$

$n = 7$ $\quad$ $\Sigma x_1 x_2 = 410.80$

$$r = \frac{410.8 - \dfrac{(31.6)(96)}{7}}{\sqrt{(7.1686)(85.4286)}} = \frac{-22.5714}{24.7468} = -.91.$$

High correlation explains the fact that either variable is equally effective in predicting cotton yield.

Set 11A

1. $H_0: P_1 = P_2 = P_3 = P_4 = P_5 = \dfrac{1}{5}$.

H_a: at least one of the above equalities is incorrect.

$E(n_i) = np_i = 250\left(\dfrac{1}{5}\right) = 50$ for $i = 1, 2, \ldots, 5$

With $k - 1 = 5 - 1 = 4$ degrees of freedom, reject H_0 if $X^2 > \chi^2_{.05} = 9.49$.

Test statistic: $X^2 = \Sigma \dfrac{[n_i - E(n_i)]^2}{E(n_i)}$

$$= \frac{(62 - 50)^2 + (48 - 50)^2 + (56 - 50)^2 + (39 - 50)^2 + (45 - 50)^2}{50}$$

$$= \frac{144 + 4 + 36 + 121 + 25}{50} = 6.6.$$

Do not reject H_0. We cannot say there is a preference for color.

2. $H_0: P_1 = .41, P_2 = .12, P_3 = .03, P_4 = .44$.

H_a: at least one equality is incorrect.

Blood Type	A	B	AB	O
Observed n_i	90	16	10	84
Expected $E(n_i)$	82	24	6	88

With $k - 1 = 3$ degrees of freedom, reject H_0 if $X^2 > \chi^2_{.05} = 7.81$.

Test statistic: $X^2 = \dfrac{(90 - 82)^2}{82} + \dfrac{(16 - 24)^2}{24} + \dfrac{(10 - 6)^2}{6} + \dfrac{(84 - 88)^2}{88}$

$$= .7805 + 2.6667 + 2.6667 + .1818 = 6.30$$

Do not reject H_0. There is insufficient evidence to refute the given proportions.

Set 11B

1. H_0: independence of classifications.

   H_a: classifications are not independent.

   Expected and observed cell counts are:

		Income		
Party Affiliation	Low	Average	High	
Republican	33 (30.57)	85 (74.77)	27 (39.66)	145
Democrat	19 (30.78)	71 (75.29)	56 (39.93)	146
Other	22 (12.65)	25 (30.94)	13 (16.41)	60
	74	181	96	351

   With $(r - 1)(c - 1) = 2(2) = 4$ degrees of freedom, reject H_0 if $X^2 > \chi^2_{.05} = 9.49$

   Test statistic: $X^2 = \dfrac{(2.43)^2}{30.57} + \dfrac{(10.23)^2}{74.77} + \ldots + \dfrac{(-3.41)^2}{16.41} = 25.61.$

   Reject H_0. There is a significant relationship between income levels and political party affiliation.

2. H_0: opinion independent of sex.

   H_a: opinion dependent upon sex.

   Expected and observed cell counts are:

		Opinion	
Sex	For	Against	
Male	114 (116.58)	60 (57.42)	174
Female	87 (84.42)	39 (41.58)	126
	201	99	300

   With $(r - 1)(c - 1) = 1$ degree of freedom and $\alpha = .05$, reject H_0 if $X^2 > \chi^2_{.05} = 3.84$. Test statistic is

   $$X^2 = \frac{(-2.58)^2}{116.58} + \frac{(2.58)^2}{57.42} + \frac{(2.58)^2}{84.42} + \frac{(-2.58)^2}{41.58} = .4119$$

   Do not reject H_0. There is insufficient evidence to show that opinion is dependent upon sex.

Set 11C

1. $H_0: p_1 = p_2 = p_3 = p_4 = p.$

   $H_a: p_i \neq p$ for at least one $i = 1, 2, 3, 4.$

	Ward				
	1	*2*	*3*	*4*	
Favor	75 (66.25)	63 (66.25)	69 (66.25)	58 (66.25)	265
Against	125 (133.75)	137 (133.75)	131 (133.75)	142 (133.75)	535
	200	200	200	200	800

With $(r - 1)(c - 1) = 3$ degrees of freedom and $\alpha = .05$, reject H_0 if $X^2 > \chi^2_{.05} = 7.81$.

Test statistic: $X^2 = \dfrac{8.75^2}{66.25} + \dfrac{(-3.25)^2}{66.25} + \ldots + \dfrac{(8.25)^2}{133.75} = \dfrac{162.75}{66.25} + \dfrac{162.75}{133.75} = 3.673.$

Do not reject H_0. There is insufficient evidence to suggest a difference from ward to ward.

2. H_0: independence of classifications.

   H_a: dependence of classifications.

	Categories			
Satisfaction	*I*	*II*	*III*	*IV*
High	40 (50)	60 (50)	52 (50)	48 (50)
Medium	103 (90)	87 (90)	82 (90)	88 (90)
Low	57 (60)	53 (60)	66 (60)	64 (60)

With $(r - 1)(c - 1) = 6$ degrees of freedom, reject H_0 if $X^2 > \chi^2_{.05} = 12.59$.

Test statistic: $X^2 = \dfrac{(-10)^2}{50} + \dfrac{10^2}{50} + \ldots + \dfrac{4^2}{60} = 8.727.$

Do not reject H_0.

Set 11D

1. With $e^{-2} = .135335$, $p(0) = .135335$, $p(1) = 2e^{-2} = .270670$, $p(2) = 2e^{-2} = .27067$ and $P[y \geq 3] = 1 - p(0) - p(1) - p(2) = .323325$. The observed and expected cell counts are shown below.

n_i	4	15	16	15
$E(n_i)$	6.77	13.53	13.53	16.17

With $k - 1 = 3$ degrees of freedom, reject H_0 if $X^2 > \chi^2_{.05} = 7.81$.

Test statistic: $X^2 = \dfrac{(-2.77)^2}{6.77} + \dfrac{(1.47)^2 + (2.47)^2}{13.53} + \dfrac{(-1.17)^2}{16.17}$

$= 1.133 + .611 + .085 = 1.829.$

Do not reject the model.

2. Expected numbers are $E(n_i) = np_i = 100 \, p_i$, where $p_i = P$ [observation falls in cell i | score drawn from the standard normal distribution]. Hence, using Table 3,

$$p_1 = P[z < -1.5] = .0668;$$

$$p_2 = P[-1.5 < z < -0.5] = .2417;$$

$$p_3 = P[-0.5 < z < 0.5] = 2\,(.1915) = .3830;$$

$$p_4 = .2417;$$

$$p_5 = .0668.$$

Observed and expected cell counts are shown below.

n_i	8	20	40	29	3
$E(n_i)$	6.68	24.17	38.30	24.17	6.68

With $k - 1 = 4$ degrees of freedom, reject H_0 if $X^2 > x^2_{.05} = 9.49$.

Test statistic: $X^2 = \dfrac{1.32^2}{6.68} + \dfrac{(-4.17)^2}{24.17} + \ldots + \dfrac{(-3.68)^2}{6.68}$

$$= .2608 + .7194 + .0755 + .9635 + 2.0273 = 4.047.$$

Do not reject the model.

Set 12A

1. Each student will obtain different results, depending on his randomization method. For example, using line 19, column 4 as a starting point, reading to the right in groups of three digits, the following 25 numbers (each corresponding to a particular tenant) are selected:

585	831	643	351	283
862	499	509	563	378
321	873	473	574	709
614	623	817	907	426
513	495	752	056	669

2. Number the experimental units from 1 to 20. Choose a starting point and direction. The first 5 two-digit numbers (between 01 and 20) will indicate the experimental units assigned to treatment 1, the second 5 to treatment 2, the third 5 to treatment 3. The remaining units are assigned to treatment 4.

3. a. No. The experimenter will inevitably catch the slower, bigger, or heavier animal. Each animal will not have an equal chance of selection.
   b. Randomly choose a one-digit number. If it is 1, 2, 3, 4, or 5 choose animal 1; if it is 6, 7, 8, 9, or 0, choose animal 2.

Set 12B

1. σ_1 = 6 and σ_2 = 4. Hence, 3/5 of the units should be assigned to population 1 and 2/5 of the units to population 2. Then,

$$n_1 = \frac{3}{5}(50) = 30 \quad \text{and} \quad n_2 = 20.$$

2. a. The ten values of x are 120, 140, 160, 180, 200, 220, 240, 260, 280, 300, with $\Sigma x_i = 2100$ and $\Sigma x_i^2 = 474000$. Then $\Sigma(x_i - \bar{x})^2 = 474000 - 441000 = 33000$ and

$$\sigma_{\hat{\beta}_1} = \frac{10}{\sqrt{33000}} = \frac{1}{\sqrt{330}}.$$

b. $\Sigma x_i = 2100$, $\Sigma x_i^2 = 522000$, $\Sigma(x_i - \bar{x})^2 = 522000 - 441000 = 81000$

$$\sigma_{\hat{\beta}_1} = \frac{10}{\sqrt{81000}} = \frac{1}{\sqrt{810}}.$$

c. $\dfrac{\sqrt{810}}{\sqrt{330}} = 1.57$ times as large.

d. Allocation b. is optimal.

Set 12C

1. a. The sales of each of the several brands will be measured in each of the three categories, using the three categories as blocks.
   b. Use twelve sheets of plywood, four of each wood type, and apply the four glues randomly to each of the three types. Here, "wood type" is the block.
2. Each treatment (A, B, C, and D) must appear once in each row and column.

D	A	C	B
B	C	A	D
C	B	D	A
A	D	B	C

3. Each block must be a homogeneous unit. Hence, blocks must be laid out from north to south.

Set 13A

1. $CM = \dfrac{(93.8)^2}{15} = 586.5627$; Total $SS = 596.26 - CM = 9.6973$.

$$SST = \frac{32.3^2 + 28.2^2 + 33.3^2}{5} - CM = 589.484 - CM = 2.9213.$$

$SSE = 9.6973 - 2.9213 = 6.7760.$

ANOVA

Source	d.f.	SS	MS	F
Treatments	2	2.9213	1.4607	2.5867
Error	12	6.7760	.5647	
Total	14	9.6973		

a. $H_0: \mu_1 = \mu_2 = \mu_3.$

H_a: at least one of the equalities is incorrect.

Test statistic: $F = \dfrac{MST}{MSE}.$

With $v_1 = 2$ and $v_2 = 12$ degrees of freedom, reject H_0 if $F > F_{.05} = 3.89$ Since $F = 2.5867$, do not reject H_0. We cannot find a significant difference.

b. $(\bar{T}_2 - \bar{T}_3) \pm t_{.025} \sqrt{MSE \left(\dfrac{1}{n_2} + \dfrac{1}{n_3} \right)}$

$= (5.64 - 6.66) \pm 2.179 \sqrt{\dfrac{2(.5647)}{5}} = -1.02 \pm 2.179\,(.4753)$

$= -1.02 \pm 1.04; \quad -2.06 < \mu_2 - \mu_3 < .02$ with 95% confidence.

2. $CM = \dfrac{(112)^2}{20} = 627.2; \quad$ Total $SS = 708 - CM = 80.80.$

$SST = \dfrac{31^2 + 17^2 + 39^2 + 25^2}{5} - CM = 679.2 - CM = 52.00.$

$SSE = 80.80 - 52.00 = 28.80.$

ANOVA

Source	d.f.	SS	MS	F
Areas	3	52.00	17.33	9.63
Error	16	28.80	1.80	
Total	19	80.80		

a. $H_0: \mu_1 = \mu_2 = \mu_3 = \mu_4;$ H_a: at least one equality does not hold.

$F = 9.63 > F_{.01} = 5.29.$ Reject H_0. There is a difference for the four areas.

b. $\bar{T}_1 \pm t_{.025} \sqrt{MSE \left(\dfrac{1}{n_1} \right)} = 6.2 \pm 2.120 \sqrt{1.80 \left(\dfrac{1}{5} \right)}$

$= 6.2 \pm 2.12\,(.6) = 6.2 \pm 1.272.$

c. $(\bar{T}_1 - \bar{T}_3) \pm t_{.025} \sqrt{MSE\left(\dfrac{1}{n_1} + \dfrac{1}{n_3}\right)} = (6.2 - 7.8) \pm 2.120 \sqrt{1.80\left(\dfrac{2}{5}\right)}$

$\qquad = -1.6 \pm 2.12\,(.85) = -1.6 \pm 1.802.$

Set 13B

1. $CM = \dfrac{(1453)^2}{20} = 105560.45;\quad$ Total $SS = 106399 - CM = 838.55.$

$SST = \dfrac{362^2 + 343^2 + 366^2 + 382^2}{5} - CM = 105714.6 - CM = 154.15.$

$SSB = \dfrac{310^2 + 270^2 + \ldots + 266^2}{4} - CM = 106050.25 - CM = 489.80.$

$SSE = 838.55 - 154.15 - 489.80 = 194.60.$

ANOVA

Source	d.f.	SS	MS	F
Litters	4	489.80	122.45	7.55
Additive	3	154.15	51.38	3.17
Error	12	194.60	16.22	
Total	19	838.55		

a. $F = 3.17 < 3.49.$ Do not reject H_0. Insufficient evidence to detect a difference due to additives.

b. $F = 7.55 > 3.26.$ Reject H_0. There is a difference due to litters. Blocking is desirable.

c. $(\bar{T}_1 - \bar{T}_2) \pm t_{.025}\sqrt{MSE\left(\dfrac{2}{b}\right)} = \dfrac{362 - 343}{5} \pm 2.179\sqrt{6.488}$

$\qquad = 3.8 \pm 5.6.$

d. $\bar{T}_3 \pm t_{.025}\sqrt{\dfrac{MSE}{b}} = \dfrac{366}{5} \pm 2.179\sqrt{\dfrac{16.22}{5}} = 73.2 \pm 3.9.$

2. $CM = \dfrac{(922)^2}{12} = 70840.3333;\quad$ Total $SS = 72432 - CM = 1591.6667.$

$SST = \dfrac{453^2 + 469^2}{6} - CM = \dfrac{425170}{6} - CM = 21.3334.$

$SSB = \dfrac{161^2 + 134^2 + \ldots + 118^2}{2} - CM = 72391 - CM = 1550.6667.$

$SSE = 19.6666.$

ANOVA

Source	d.f.	SS	MS	F
Treatments	1	21.3334	21.3334	5.42
Blocks	5	1550.6667	310.1333	78.85
Error	5	19.6666	3.9333	
Total	11	1591.6667		

b. $F = 5.42 = t^2 = (2.328)^2$.

Set 13C

1. $CM = \dfrac{705^2}{25} = 19881$; Total $SS = 21321 - CM = 1440.0$.

$$SSR = \frac{161^2 + 110^2 + \ldots + 108^2}{5} - CM = 20680.2 - CM = 799.2$$

$$SSC = \frac{122^2 + \ldots + 165^2}{5} - CM = 20099 - CM = 218.0.$$

$$SST = \frac{121^2 + 163^2 + 162^2 + 130^2 + 129^2}{5} - CM = 318.0.$$

a.

ANOVA

Source	d.f.	SS	MS	F
Rows	4	799.2	199.8	22.89
Columns	4	218.0	54.5	6.24
Treatments	4	318.0	79.5	9.11
Error	12	104.8	8.7	
Total	24	1440.0		

b. $F = 9.11 > F_{.05} = 3.26$. There is a difference in the mean yields for the five varieties of corn.

c. $F_R = 22.89$ and $F_C = 6.24$ are both greater than 3.26. Rows and columns are both significant.

d. Columns are significant and should not be eliminated from the design.

e. $(\bar{T}_B - \bar{T}_E) \pm t_{.025} \sqrt{MSE\left(\dfrac{2}{p}\right)} = \dfrac{163 - 129}{5} \pm 2.179\sqrt{3.48}$

$\qquad = 6.8 \pm 4.07$.

2. $CM = \dfrac{1629^2}{16} = 165852.5625$; Total $SS = 167681 - CM = 1828.4375$.

$$SSR = \frac{665927}{4} - CM = 629.1875; \quad SSC = \frac{664325}{4} - CM = 228.6875.$$

$$SST = \frac{447^2 + 424^2 + 388^2 + 370^2}{4} - CM = 166757.25 - CM = 904.6875.$$

a.

ANOVA

Source	d.f.	SS	MS	F
Days	3	629.1875	209.7292	19.10
Suppliers	3	228.6875	76.2292	6.94
Treatments	3	904.6875	301.5625	27.47
Error	6	65.8750	10.9792	
Total	15	1828.4375		

b. $F = 27.47 > 9.78$. There is a significant difference among methods.

c. $(\bar{T}_A - \bar{T}_B) \pm t_{.025} \sqrt{\dfrac{2MSE}{4}} = \dfrac{447 - 424}{4} \pm 2.447 \sqrt{5.4896}$

$= 5.75 \pm 5.73.$

d. $F = 6.94 > 4.76$. There is a significant difference among suppliers.

e. $(\bar{C}_1 - \bar{C}_2) \pm t_{.025} \sqrt{\dfrac{2MSE}{4}} = \dfrac{399 - 432}{4} \pm 5.73 = -8.25 \pm 5.73.$

f. $F = 19.10 > 9.78$. Blocking on days is effective.

Set 14A

1. Let $p = P$ [paint A shows less wear] and $y =$ number of locations where paint A shows less wear. Since no numerical measure of a response is given, the sign test is appropriate.

$H_0: p = \frac{1}{2}.$

$H_a: p < \frac{1}{2}.$

Rejection region: With $n = 25$, $p = \dfrac{1}{2}$, reject H_0 if $y \leqslant 8$ with $\alpha = .054$. (See Table 1 e.)

Observe $y = 8$; therefore, reject H_0. Paint B is more durable.

2. Rank the times from low to high. Note $n_1 = n_2 = 5$.

Water	Food
2	6
7	8
3	4
1	10
5	9

$H_0:$ no difference in the distributions.

$H_a:$ the distributions are different.

Rejection region: Use the convention of choosing the smaller of U and U', so that we are only concerned with the lower portion of the rejection region. Using Table 8, reject H_0 if the smaller of U and U' is less than or equal to 3 with $\alpha/2 = .0278$, so that $\alpha = .0556$.

Calculate $U = 5(5) + \dfrac{1}{2}(5)(6) - 18 = 22.$

$$U' = 5(5) + \dfrac{1}{2}(5)(6) - 37 = 3. \text{ Reject } H_0.$$

3. $H_0: p = \dfrac{1}{2}$, where $p = P$ [Variety A exceeds Variety B].

$H_a: p \neq \dfrac{1}{2}$.

With $\alpha = .022$ from Table 1, reject H_0 if $y = 0, 1, 9, 10$. Since y = number of plus signs = 8, do not reject H_0. We cannot detect a difference between varieties A and B.

4. Rank the scores from low to high. Let $n_1 = 7, n_2 = 8.$

Women (1)	Men (2)	
6 (10)	3 (6)	H_0: No difference in the distributions.
10 (15)	5 (9)	
3 (6)	2 (4)	H_a: Scores are higher for women.
8 (12.5)	0 (1.5)	
8 (12.5)	3 (6)	
7 (11)	1 (3)	
9 (14)	0 (1.5)	
	4 (8)	

Rejection region: Women should have higher ranks if H_a is true, making $U = n_1 n_2 + \dfrac{1}{2} n_1 (n_1 + 1) - T_1$ small. Using Table 8, reject H_0 if $U \leq 13$ with $\alpha = .0469.$

Calculate $U = 7(8) + \dfrac{1}{2}(7)(8) - 81 = 3.$ Reject $H_0.$

Set 14B

1. H_0: No difference in distributions of number of nematodes for varieties A and B.

H_a: Distribution of number of nematodes differs for varieties A and B.

Rank the absolute differences from smallest to largest and calculate T, the smaller of the two (positive and negative) rank sums.

Location	1	2	3	4	5	6	7	8	9	10
d_i	186	138	349	120	-8	92	219	39	-32	52
Rank $|d_i|$	8	7	10	6	1	5	9	3	2	4

Rejection region: With $\alpha = .02$ and a two-sided test, reject H_0 if $T \leq 5$. Since $T = 1 + 2 = 3$, reject H_0. There is a difference between A and B.

2. $H_0: p = \dfrac{1}{2}$, where $p = P$ [positive difference] and $n = 24.$

$$H_a : p > \frac{1}{2}.$$

Using $\alpha = .05$ and the normal approximation, H_0 will be rejected if

$$\frac{y - .5n}{.5\sqrt{n}} > 1.645,$$

where $y =$ number of positive differences. Calculate

$$z = \frac{y - 12}{.5\sqrt{24}} = \frac{16 - 12}{2.45} = 1.63. \text{ Do not reject } H_0.$$

3. The ranks of the absolute differences are given below along with their corresponding signs.

-12, 13, 6, -4, -10, 3, 14.5, 20, 8, 16, -9, -1, 22, 17, 11, 18, 19, 24, -5, 21, -7, 14.5, -2, 23.

With $\alpha = .05$ and a one-sided test, reject H_0 if $T \leqslant 92$. Since $T = 50$, reject H_0. Note that the sign test is computationally simple, but the Wilcoxon test is more efficient since it allows us to reject H_0 while the sign test did not.

Set 14C

1. H_0: Advertising has no effect on sales.
   H_a: Advertising increases sales of advertised model.

   Rejection region: With $n_1 = 13, n_2 = 14$ the normal approximation is used and H_0 is rejected if $z = \frac{R - E(R)}{\sigma_R} < -1.645.$

   Calculate $E(R) = 1 + \frac{2(13)(14)}{27} = 14.48.$

   $$\sigma_R^2 = \frac{364(337)}{27^2(26)} = \frac{122668}{18954} = 6.4719.$$

   $$z = \frac{9 - 14.48}{\sqrt{6.4719}} = \frac{-5.48}{2.54} = -2.16. \text{ Reject } H_0.$$

2. The sequence of runs is

   M M M M M M W M M W W W W W W,

   or   M M M M W M M M M W W W W W W,

   or   M M M M M W M M M W W W W W W,

   depending on how the three observations with rank 6 are arranged. In any case, $R = 4$. The rejection region, with $n_1 = 7, n_2 = 8$, and $\alpha = .051$ is to reject H_0 if $R \leqslant 5$. Hence, H_0 is rejected and we conclude that status-frustration scores are higher among women.

3. Rank the examination scores, and note that the interview scores are already in rank order.

Interview Rank (x_i)	Exam Rank (y_i)
4	3
7	2
5	5
3	4
6	1
2	6
1	7

$n = 7$ $\qquad\qquad$ $\Sigma x_i^2 = 140$

$\Sigma x_i = 28$ $\qquad\qquad$ $\Sigma y_i^2 = 140$

$\Sigma y_i = 28$ $\qquad\qquad$ $\Sigma x_i y_i = 88$

$$r_s = \frac{7(88) - (28)^2}{7(140) - (28)^2} = \frac{616 - 784}{980 - 784} = -.857.$$

To test $H_0: \rho_s = 0$; $H_a: \rho_s < 0$, the rejection region is $r_s < -.714$.
Hence, H_0 is rejected with $\alpha = .05$.

4. a. $\Sigma x_i = \Sigma y_i = 45$.

$\Sigma x_i^2 = \Sigma y_i^2 = 285$

$\Sigma x_i y_i = 272$

$$r_s = \frac{9(272) - (45)^2}{9(285) - (45)^2} = \frac{423}{540} = .78.$$

b. $H_0: \rho_s = 0$.

$H_a: \rho_s > 0$.

With $\alpha = .05$, reject H_0 if $r_s > .600$. Reject H_0. They are in basic agreement.

ANSWERS TO EXERCISES

Chapter 2

1. $u.$
2. $3a^3 - 2a.$
3. 2; 2/9.
4. $2/x^2 - 3/x + 5.$
5. $2/v^2 - 3v.$
6. $x_3 + x_4 + x_5 + x_6 - 4a.$
7. $\displaystyle\sum_{i=3}^{6} (x_i - m)^2.$
8. $9x + 9.$
9. 1.
10. 60.
11. 25.
12. 1750.

Chapter 3

1. a. Range = 6.8. c. .80; .55. e. Median = 19.5. f. 10.
   g. 75 (upper quartile). h. $\bar{y} = 19.03$; $s^2 = 2.7937$; $s = 1.67$.
   i. Yes, since 70%, 95% and 100% of the measurements lie in the intervals
   $\bar{y} \pm ks$, $k = 1, 2, 3$, respectively. j. Yes (see i).
2. a. 16%. b. 81.5%.
3. a. 8. b. 2.45 to 2.95; 5.95 to 6.45.
4. $\mu = 6.6$ oz.
5. a. s is approximated as 16. b. $\bar{y} = 136.07$; $s^2 = 292.4952$; $s = 17.1$.
   c. $a = 101.82$; $b = 170.27$. d. Yes, for approximate calculations. e. No.
7. a. s is approximated as 5. b. $\bar{y} = 11.67$; $s^2 = 13.9523$; $s = 3.74$.
   c. 14/15.
8. a. Approximately 97.4%. b. Approximately 16%.
9. Approximately .025.

Chapter 4

1. a. $(ABC), (ACB)$. b. $(CAB), (ACB)$. c. $(ABC), (ACB), (CAB)$. d. $(AC$
   e. $P(A) = 1/3$; $P(A|B) = 1/2$; A and B are dependent.
2. a. 1/2. b. 1/2. c. 5/6.
3. .1792.
4. a. 5/32. b. 31/32.
5.

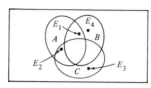

6. a. 1/6. b. 2/3; 1/3. c. 2/3; 2/3. d. 5/6; 5/6. e. no; no.
7. a. 56. b. 30. c. 15/28.
8. .41.
9. a. .328. b. .263.
10. .045; yes.
11. a. 14/22; 4/22; 4/22. b. 35/44; 8/44; 1/44.
12. a. 99/991. b. 892/991. c. 3% defective.

Chapter 5

1.

y	$p(y)$
0	.1
1	.6
2	.3

2. a.

y	$p(y)$
0	8/27
1	12/27
2	6/27
3	1/27

3. $E(y) = 11250$; $\sigma = 5673.4$.
4. $E(R) = 112,500$; $\sigma_R = 56734.0$.
5. .65.
6. a. $E(y) = 2$; $\sigma^2 = 1$.
7. a. 1/4. b. 3/16. c. 9/64. d. $p(y) = \left(\dfrac{3}{4}\right)^{y-1} \left(\dfrac{1}{4}\right)$ e. yes.
8. $6.50.
9. $2.70.

Chapter 6

1. .655360.

. a. $p(y) = \dfrac{20!}{y!\,(20-y)!}\,(.3)^y\,(.7)^{20-y}$ $y = 0, 1, 2, \ldots, 20$.

  b. .772. c. .780.

. a. 82. b. 76.

. a. .9801. b. .6400.

. a. $a = 1$. b. .069. c. P [accept lot | p low] is higher and P [reject lot | p high] is lower; more costly.

. $\alpha = .001$, $\beta = .05$.

. a. $H_0: p = .4$. b. .166. c. $H_a: p > .4$. d. .367.

. a. Declaration that the cheeses differ when they are equally desirable.

  b. Failure to declare that a difference exists when one of the cheeses is more desirable. c. .022. d. .952. e. .264. f. .004. g. When p is close to the value specified in H_0.

. Reject H_0 for $y \geq 5$ with $\alpha = .043$. Do not reject $H_0: p = .1$.

. Reject H_0 for $y \geq 8$ with $\alpha = .032$. Reject $H_0: p = .2$. The drug is effective.

Chapter 7

1. a. .9713. b. .1009. c. .7257. d. .9706. e. .8925. f. .5917.
2. a. $z_0 = .70$. b. $z_0 = 2.13$. c. $z_0 = 1.645$. d. $z_0 = 1.55$.
3. a. .8413. d. .8944. c. .9876. d. .0401.
4. no.
5. a. .9838. b. .0000. c. .8686.
6. .3520.
7. 87.48.
8. $\mu = 10.071$.
9. a. .0139. b. .0668. c. .5764. d. .00000269.
0. a. .1635. b. .0192. c. Yes, since $P\,[y \geq 60 \mid p = .2] = .0192$.

Chapter 8

1. The inference; measure of goodness.
2. Unbiasedness; minimum variance.
3. 61.23 ± 1.50.
4. $2.705 \pm .012$.
5. $.030 \pm .033$.
7. $z = 2.5$; yes.
8. $.6 \pm .048$.
9. Approximately 256.
10. Approximately 100.
11. Approximately 100.
12. -8 ± 4.49.
13. $z = -5.2$; reject the claim.
14. $z = -3.40$; yes.
15. a. $z = 5.8$; yes. b. $3.0 \pm .85$.

Chapter 9

1. According to the Central Limit Theorem, these statistics will be approximately normally distributed for large n.
2. i. The parent population has a normal distribution.
   ii. The sample is a random sample.
3. The number of degrees of freedom associated with a t-statistic is the denominator of the estimator of σ^2.
4. Do not reject H_0; $t = -.6$.
5. $2.48 < \mu < 4.92$.
6. Do not reject H_0; $t = 1.16$.
7. $-0.76 < \mu_1 - \mu_2 < 3.96$.
8. Do not reject H_0; $t = 2.29$.
9. Do not reject H_0; $t = 1.48$.
10. Do not reject H_0; $\chi^2 = 8.19$.
11. $.214 < \sigma^2 < 4.387$.
12. Do not reject H_0; $F = 1.796$.
13. $.565 < \dfrac{\sigma_1^2}{\sigma_2^2} < 5.711$.
14. Reject H_0; $F = 2.06$.

Chapter 10

1.

	y-Intercept	Slope
a.	-2	3
b.	0	2
c.	-0.5	-1
d.	2.5	-1.5
e.	2	0

2. a. $\hat{y} = 2 - .875\,x$. b. $SSE = .25$; $s^2 = .0833$; $s = .289$.
   d. Reject H_0; $\beta_1 = 0$, since $t = 12.12$.
   e. $2.525 < E(y|x = -1) < 3.225$.
   f. $r^2 = .98$. The use of the linear model rather than $\bar{y}$ as a predictor for y reduced the sum of squares for error by 98%.
   g. $-.345 < y_p < 2.595$. h. $\bar{x}$.
3. The fitted line may not adequately describe the relationship between x and y outside the experimental region.
4. The error will be a maximum for the values of x at the extremes of the experimental region.
5. If $r = 1$, the observed points all lie on the fitted line having a positive slope.
   If $r = -1$, the observed points all lie on the fitted line having a negative slope.
6. a. $\hat{y} = 7.0 + 15.4x$. b. $SSE = 50.4$; $s^2 = 8.4$.
   c. Reject $H_0: \beta_1 = 0$, since $t = 16.7$.
   d. $43.0 < E(y|x = 2.5) < 48.0$
   e. $13.6 < \beta_1 < 17.2$. f. $r^2 = .979$. (See Exercise 2f.)

7. a. $\hat{y} = 6.96 + 2.31\,x$.   c. $SSE = .9751$; $s^2 = .1219$.
   d. Reject $H_0: \beta_1 = 0$, since $t = 19.25$.
   e. $r = .99$.   f. $r^2 = .979$.
   g. Do not reject H_0; $t = 1.67$.
   h. $9.27 \pm .69$; $8.58 < y_p < 9.96$.
8. a. $\hat{y} = 20.47 - .76\,x$.   b. $SSE = 4.658$; $s^2 = .5822$.
   c. Reject H_0; $t = -22.3$.   d. $-.86 < \beta_1 < -.66$.
   e. $9.83 \pm .55$; $9.28 < E(y|x = 14) < 10.38$.
   f. $r^2 = .984$. (See Exercise 2f.)

Chapter 11

2. a. Yes.   b. No, since p_i, $i = 1, 2, 3$ changes from trial to trial.
   c. Yes.
3. a. 8.4.   b. 18.0.   c. 75.6.
4. Reject H_0, $X^2 = 16.535$.
5. Do not reject H_0, $X^2 = 2.300$.
6. Reject H_0, $X^2 = 7.97$.
7. Do not reject H_0: $p_1 = p_2 = p_3 = p_4$, $X^2 = 1.709$.
8. Reject H_0, $X^2 = 9.333$.
9. Do not reject the model, $X^2 = 6.156$.

Chapter 12

1. The smaller the confidence interval, the more precise is the estimate of a population parameter, and hence the more information in a sample.
2. Noise refers to the variation in the data, while volume refers to information pertinent to one or more population parameters.
3. An object upon which a measurement is taken is called an experimental unit.
4. A factor is an independent experimental variable.
5. A quantitative factor is one that can take values corresponding to points on the real line. A qualitative variable is one that is not quantitative.
6. A treatment is a specific combination of factor levels.
7. See Section 12.3.
8. See Section 12.4.
9. See Section 12.4.
2. Yes. One or two points should be run at the center of the design to detect curvature in the response.

Chapter 13

1. a. Completely randomized design.
   b.

Source	d.f.	SS	MS	F
Chemicals	2	25.1667	12.5834	2.59
Error	9	43.75	4.8611	
Total	11	68.9167		

c. No.   d. -2.25 ± 3.53 e. 11 ± 2.02.

f. 39.

2. a. Randomized block design.

b.

Source	d.f.	SS	MS	F
Applications	3	18.9167	6.3056	9.87
Chemicals	2	62.1667	31.0833	48.65
Error	6	3.8333	0.6388	
Total	11	84.9167		

c. $F = 48.65$; reject H_0; yes.   d. 5.25 ± 1.38.   e. 21.   f. Yes.

3. a. Completely randomized design.

b.

Source	d.f.	SS	MS	F
Treatments	3	1052.68	350.89	1.76
Error	15	2997.95	199.86	
Total	18	4050.63		

c. $F = 1.76$; no.

d. -12.6 ± 19.06.

4. a.

Source	d.f.	SS	MS	F
Programs	2	25817.49	12908.74	4.43
Error	13	37851.51	2911.65	
Total	15			

$F = 4.43$; yes.

b. -63.1 ± 59.9.

5. a.

Source	d.f.	SS	MS	F
Typists	3	198.5	66.17	3.04
Periods	3	119.5	39.83	1.83
Brands	3	696.5	232.17	10.67
Error	6	130.5	21.75	
Total	15			

b. $F = 10.67$; yes.

c. $F = 3.04$; no.

d. $F = 1.83$; no.

e. 5.75 ± 8.07.

6. a.

ANOVA				
Source	d.f.	SS	MS	F
Employees	11	477.8889	43.4444	2.9444
Conditions	2	230.7222	115.3611	7.8184
Error	22	324.6111	14.7551	
Total	35			

$F = 7.8184 > 3.44$. Reject H_0. There is a significant difference between conditions.

b. $F = 2.944$; yes.

c. $-9.34 < \mu_2 - \mu_3 < 2.84$.

Chapter 14

1. Sign test: both; Mann-Whitney U-test: independent; Wilcoxon test: related; runs test: independent.

2. $z = -2.54 < -1.645$. Reject H_0.

3. Reject H_0 when $U \leqslant 28$. $U = 16$. Reject H_0.

4. Reject H_0 when $R \leqslant 7$. $R = 10$. Do not reject H_0.
The runs test is less efficient in detecting a difference in population means.

5. $r_s = 1$. While $r = 1$ only when the data points all lie on the same straight line, r_s will be 1 whenever y increases steadily with x.

6. a. $.738$. b. $.738 \geqslant .643$; reject H_0.

7. With $\alpha = .032$, reject H_0 when $R \geqslant 13$. $R = 13$. Reject H_0.

8. With $\alpha = .0703$, reject H_0 if $y = 0, 1, 7, 8$. $y = 6$. Do not reject H_0.

9. Reject H_0 if $T \leqslant 6$, $\alpha = .10$. $T = 9.5$. Do not reject H_0.

10. Reject H_0 if $U \leqslant 21$ with $\alpha = .0506$. $U = 15.5$. Reject H_0.
Operator A produces a lower percentage defective.